# 1st International Conference on Recent Advances in Aerospace Engineering (ICRAAE-2017)



IOP Conference Series: Materials Science and Engineering
Volume 247

Coimbatore, India
3 – 4 March 2017

ISBN: 978-1-5108-4995-2
ISSN: 1757-8981

**Printed from e-media with permission by:**

Curran Associates, Inc.
57 Morehouse Lane
Red Hook, NY 12571



Printed by Curran Associates, Inc. (2017)

For permission requests, please contact the Institute of Physics at the address below.

Institute of Physics
Dirac House, Temple Back
Bristol BS1 6BE UK

Phone: 44 1 17 929 7481
Fax: 44 1 17 920 0979

techtracking@iop.org

**Additional copies of this publication are available from:**

Curran Associates, Inc.
57 Morehouse Lane
Red Hook, NY 12571 USA
Phone: 845-758-0400
Fax: 845-758-2633
Email: curran@proceedings.com
Web: www.proceedings.com

# Table of contents

## Volume 247

**1st International Conference on Recent Advances in Aerospace Engineering (ICRAAE-2017)**

**3–4 March 2017, Coimbatore, India**

**Accepted papers received: 25 September 2017**
**Published online: 16 October 2017**

### Preface



### Shock Waves

## Control and Instrumentation



## Aerodynamics

## Materials

IOP Conf. Series: Materials Science and Engineering **247** (2017) 011001 doi:10.1088/1757-899X/247/1/011001

# FIRST INTERNATIONAL CONFERENCE ON RECENT ADVANCES IN AEROSPACE ENGINEERING (ICRAAE-2017)

**3rd and 4th March, 2017**

**Karunya Institute of Technology and Sciences**
**Karunya University,**
**(Declared as Deemed to be University under sec. 3 of the UGC Act, 1956)**
**coimbatore, TamilNadu , India.**

## Introduction

The First International conference on Recent Advances in Aerospace Engineering (ICRAAE-2017) will be conducted by the Department of Aerospace Engineering at the Karunya University, Coimbatore, Tamilnadu, India, on 3rd and 4th March, 2017. The conference aims to bring together students, academicians, leading scientists, researchers and industrialists working in the diverse fields of Aerospace Engineering. This conference provides an inter-disciplinary platform for the educators, researchers and practitioners to present, share and discuss the recent trends, innovations, concerns and solutions in the cutting edge technologies of Aerospace Engineering for mutual benefit and the growth of the nation.

## Objectives

The conference is devoted to benefit the participants who will have the opportunity to gain insight into the state-of-the art technologies in the field of Aerospace Engineering by the expert lectures of scientists and pioneering researchers from abroad and India. In addition, the two-day conference will enable knowledge sharing by personnel involved in active research working on the recent developments in this diverse field.

## Target Group

The conference welcomes students and Researchers all over the world, Academic Faculty and Scientists, Industrialists and Practitioners, from India and other nations who are associated with Aerospace Engineering and its allied disciplines

IOP Conf. Series: Materials Science and Engineering **247** (2017) 011001 doi:10.1088/1757-899X/247/1/011001

**Major Themes**

- Space Dynamics and Interplanetary trajectories
- Experimental and computational aerodynamics
- Aerospace Propulsion
- Materials and Composites for Aerospace applications
- Communication, Navigation, Guidance and Control

**Sub Themes**

- Flight Mechanics
- Rocket Propulsion & Cryogenic Propulsion
- Structural analysis
- Future Space Missions / Enabling Technologies
- Antenna, RF/Microwave Systems and Technologies
- Spacecraft Design and Systems Engineering
- Air Vehicle Systems and Technologies
- Spacecraft/Launch Vehicle Systems and Technologies
- Mechatronics, Robotics and Intelligent Systems
- Human-systems Interaction
- Modeling and Simulation Technologies
- Wind Tunnel Testing
- Atmospheric and Space Environments
- Aeroelasticity and Structural Vibration
- Rocketry and Space
- Flight Acoustics
- Manufacturing Techniques
- Defense and Security
- Aerospace Operations and Management
- Government Plans, Policies and Education
- Avionics
- Flight dynamics, Stability and Control
- Computational Flow analysis
- Thermal analysis and Heat Exchangers
- Re-entry vehicles design and analysis
- UAV and MAV design
- Aerostats and Airships
- Corrosion and Aerospace Components
- Hypersonic Flows
- Analysis and Optimization in Aerospace Engineering
- Biomimetics in Aerospace
- Life estimation of Aircraft components and Systems

IOP Conf. Series: Materials Science and Engineering **247** (2017) 011001 doi:10.1088/1757-899X/247/1/011001

## COMMITTEE MEMBERS

### Chief Patron

Dr. Paul Dhinakaran, Chancellor, Karunya University

### Patrons

Dr. Jeyakumar Daniel, Chancellor's Representative

Dr. S.Sundar Manoharan Vice-Chancellor,Karunya University

Dr. M. J. Xavier, Pro-Vice-Chancellor, Karunya University

Dr. C. Joseph Kennedy, Registrar, Karunya University

### Chairman

Dr. G. Prince Arulraj, Director – Engineering and Technology

### Convenor

Dr. P.D. Arumairaj, Head of the Department

### Advisors

Dr. E. J. James, Director – International Affairs, Karunya University

Dr. Pradeep Kumar, Professor, Aerospace Engineering

Dr. Ramkrishan Sharma, Professor, Aerospace Engineering

Shri. A. N. Subash, Consultant, Aerospace Engineering

### Organizing Secretaries

Dr. G. Jims John Wessley, AP(SG), Aero

Dr. R. Renjith Singh, AP, Aero

### Organizing Committee

Mr. Kalakanda Alfred Sunny, AP, Aero

Mr. Aldin Justin Sundarraj, AP, Aero

Mr. Prashant Kumar, AP, Aero

Lt. S. R. Musica, AP, Aero

Mr. M. Gopalsamy, AP, Aero

Mrs. S. Parameshwari, AP, Aero

Mrs. R. Gayathri, AP, Aero

Mr. A. Daniel Antony, AP, Aero

Mr. S. Saravanakumar, Technical Assistant, Aero

Mr. V. kumar, Technical Assistant, Aero

Mr. Y. Selvin Raja, Administrative Assistant, Aero

**International Advisory / Technical Committee**

Prof. Pini Gurfil, Technion-Israel Institute of Technology, Israel

Dr. Lee Chih Fang, University of Science, Malaysia

Dr. Prasad. S. Thankabail, United States Geological Survey, USA

Dr. Pardini Carmen, Space Flight dynamics Laboratory, Italy

Dr. Somchai Wongwises, King Mongkut's University of Technology, Thailand

Dr. Cristina Barrado Muxi, Polytechnic University of Catalonia, Spain

Dr. Ahmet Selim Dalkilic, Yildiz Technical University, Turkey

Dr. Joseph Gnanaraj, University of Connecticut, Storrs, USA

Dr. Alberto Coronas, Universitat Rovira, Spain

Dr. Rajasekaran, Jazan University, Saudi Arabia

Dr. Suthagar John, Ibri College of Technology, Sultanate of Oman

Dr. Raj Das, University of Auckland, New Zealand

Dr. S. J. Vijay, Jeonju University, Korea

Dr. I. Dinaharan, University of Johannesburg, South Africa

Dr. A. Brusly Solomon,University Van Pretoria, South Africa

**National Advisory / Technical Committee**

Dr. M.A. Atmanand,Chair,IEEE Madras section

Mr. HR Mohan,Vice Chair,IEEE Madras section

Dr. N. Kumarappan,IEEE Madras section

Dr. P. Sakthivel,IEEE Madras section

Dr. A. Nagappan,Chapter Chair,IEEE Aerospace and Electronic Systems Society

Dr. G. Rajesh, Indian Institute of Technology-Madras, Chennai

Dr. K. P. J. Reddy, Indian Institute of Science, Bangalore

Dr. K.Senthil Kumar, Madras Institute of Technology, Chennai

Shri. Sudhir Kumar Mishra, Brahmos Aerospace, Bangalore

Lt. Cdr. John Livingstone, Johnnette Technologies Pvt Ltd, Noida

Dr. Edwin Raja Dhas, Noorul Islam University, Nagercoil

Prof. Anu Jacob Paul, North Malabar Institute of Technology, Kerala

Dr. Jagadeesh, Indian Institute of Science, Bangalore

Dr. Abhishek, Indian Institute of Technology, Kanpur

Dr. J. Bruce Ralphin Rose, Regional Campus, Anna University, Tirunelveli

Dr. V. Ashok, Vikram Sarabhai Space Centre, ISRO, Trivandrum

Shri.P. Jeevanandam, Aeronautical Development Agency, Bangalore

Dr. T. Michael N Kumar, Bharathiyar College of Engg and Tech, Karaikal

Dr. K. Sundararaj, Sri Ramakrishna Engineering College, Coimbatore

Dr. Noor Danish Ahrar Mundari, Eternal University, Himachal Pradesh

Dr. Arnalt Stalin, Bharat Heavy Electricals Limited, Ranipet

Dr. G. Ramesh, Jain University, Bangalore

Dr. P. Manjith Singh, Terminal Ballistics Research Laboratory, Chandigarh

Dr. AP Haran, Park College of Engg and Tech, Coimbatore

Dr. Manjunath, National Aerospace Laboratories, Bangalore

Dr. K . Theerthamalai, Defence Research & Development Lab, Hyderabad

Prof. A K Anil Kumar, Vikram Sarabhai Space Centre, ISRO, Trivandrum

Dr. A. K. Verma, Amity University, Noida, Uttar Pradesh

Dr. B C Pillai, Professor Emeritus, Karunya University

Prof. G. Biswas, Indian Institute of Technology, Guwahati

Dr. J. K. Prasad, Deoghar campus of Birla Institute of Technology, Jharkhand

Dr. Rajeev Agarwal, Sri Aurobindo College, University of Delhi, New Delhi

Dr. G. Manikandan, SNS College of Engineering, Coimbatore

Dr. Parag Deshpande, National Aerospace Laboratories, Bangalore

Shri. S. Pandian, Vikram Sarabhai Space Centre, ISRO, Trivandrum

Dr. J. Arivudainambi, Adhiyamaan College of Engineering, Hosur

Dr. K. Ramachandran, Bharathiyar University, Coimbatore

Dr. M. Ramji, Indian Institute of Technology, Hyderabad

Dr. S. Subha Rani, PSG College of Technology, Coimbatore

Dr. B. Vinod, Robotics and Automation, PSG College of Technology, Coimbatore

Dr. G. S. R. Emil Selvan, Thiagarajar College of Engineering, Madurai

Dr. A. Salih, Indian Institute of Space Science and Technology, Trivandrum

**www.karunya.edu/aero/icraae** **icraae@karunya.edu**

IOP Conf. Series: Materials Science and Engineering **247** (2017) 011002 doi:10.1088/1757-899X/247/1/011002

**IEEE Sponsored**

**International Conference**

**on**

# RECENT ADVANCES IN AEROSPACE ENGINEERING (ICRAAE-2017)

IEEE
Advancing Technology for Humanity

Organized by

**Department of Aerospace Engineering**

**Karunya UNIVERSITY**

Coimbatore, Tamilnadu, India – 641 114

**3rd & 4th March 2017**

IOP Conf. Series: Materials Science and Engineering **247** (2017) 011002 doi:10.1088/1757-899X/247/1/011002

[redacted]

## Introduction

The First International conference on Recent Advances in Aerospace Engineering (ICRAAE-2017) will be conducted by the Department of Aerospace Engineering at Karunya University, Coimbatore, Tamilnadu, India, on 3rd and 4th March, 2017. The conference aims to bring together students, academicians, leading scientists, researchers and industrialists working in diverse fields of Aerospace Engineering. This conference provides an inter-disciplinary platform for the educators, researchers and practitioners to present, share and discuss the recent trends, innovations, concerns and solutions in the cutting edge technologies of Aerospace Engineering for mutual benefit and the growth of the nation.

## Objectives

The conference is devoted to benefit the participants who will have the opportunity to gain insight into state-of-the-art technologies in the field of Aerospace Engineering by the expert lectures of scientists and pioneering researchers from India and abroad. In addition, the two-day conference will enable knowledge sharing by personnel involved in active research working on the recent developments in this diverse field.

## Themes of the conference

***Major Themes***

* Space Dynamics and Interplanetary trajectories
* Experimental and computational aerodynamics
* Aerospace Propulsion
* Materials and Composites for Aerospace
* Communication, Navigation, Guidance and Control

### Important Dates

| | |
|---|---|
| Call for Papers | : 01-09-2016 |
| Full paper submission | : 31-12-2016 |
| Feedback and Comments | : 20-01-2017 |
| Final acceptance and Notification | : 01-02-2017 |
| Registration | : 03-02-2017 |
| Dates of Conference | : 03 & 04-03-2017 |

***Sub Themes***

* Flight Mechanics
* Structural analysis
* Antenna, RF/Microwave Systems and Technologies
* Remote Sensing / GIS / Signal and Image Processing
* Air Vehicle Systems and Technologies
* Mechatronics, Robotics and Intelligent Systems
* Modeling and Simulation Technologies
* Atmospheric and Space Environments
* Rockery and Space
* Manufacturing Techniques
* Aerospace Operations and Management
* Avionics
* Computational Flow analysis
* Re-entry vehicles design and analysis
* Aerostats and Airships
* Hypersonic Flows
* Biomimetics in Aerospace
* Rocket Propulsion & Cryogenic Propulsion
* Future Space Missions / Enabling Technologies
* Aerospace Transportation
* Spacecraft Design and Systems Engineering
* Spacecraft/Launch Vehicle Systems and Technologies
* Human-systems Interaction
* Wind Tunnel Testing
* Aeroelasticity and Structural Vibration
* Flight Acoustics
* Defense and Security
* Government Plans, Policies and Education
* Flight dynamics, Stability and Control
* Thermal analysis and Heat Exchangers
* UAV and MAV design
* Corrosion and Aerospace Components
* Analysis and Optimization in Aerospace Engineering
* Life estimation of Aircraft components and Systems

and other related topics

## Target Group

The conference welcomes

* Students and Researchers
* Academic Faculty and Scientists
* Industrialists and Practitioners from India and other nations associated with Aerospace Engineering and its allied disciplines.

www.karunya.edu/aero/icraae

## Paper Submission

Original manuscripts that are not presently under consideration for publication elsewhere and not published so far can be submitted in the IEEE format available in the conference link. The manuscripts must be in English only. The maximum length of the full paper should not exceed 6 pages. The full length paper following the guidelines should be submitted as per the procedure detailed in the conference weblink. The papers on submission will be reviewed by competent reviewers and the acceptance / revision will be communicated to the authors. Manuscripts that are not modified according to the review comments will be summarily rejected. Once a final decision is arrived, the authors may proceed for registration. It is mandatory that atleast one author must register per paper and he/she must present the paper during the conference in person.

[redacted]

IOP Conf. Series: Materials Science and Engineering **247** (2017) 011002 doi:10.1088/1757-899X/247/1/011002

### Publications

All the abstracts of registered and presented papers of ICRAAE 2017 will be published in IEEE conference proceedings with ISBN number and also submitted to IEEE Xplore for possible publication. Extended version of accepted and registered papers may be published in any one of the following Journals based on the willingness of the authors and proper modification of the format and payment of an additional nominal publication charges, as applicable.

1. **IEEE Transactions on Aerospace and Electronic Systems** (Scopus Indexed ISSN : 0018-9251 – Anna University Annexure I)
2. **IOP Conference Series: Materials Science and Engineering** (Scopus Indexed ISSN : 1757-899X)
3. **Applied Mechanics and Materials** (Indexed in SCImago Journal & Country Rank (SJR) ISSN: 1662-7482)
4. **International Journal of Vehicle Structures & Systems** (Scopus Indexed ISSN : 0975-3540 - Anna University Annexure – II)
5. **Journal of Advances in Mechanical Engineering and Science** (Indexed in Crossref & Googlescholar ISSN : 2455-0957)

*Few more to be added...*

### Registration Fee

| Category of Participation | Indian Delegates | | Foreign Delegates | |
|---|---|---|---|---|
| | IEEE Member | Non-IEEE Member | IEEE Member | Non-IEEE Member |
| Students | ₹ 3500 | ₹ 4000 | $ 300 | $ 350 |
| Research Scholars | ₹ 4500 | ₹ 5000 | $ 350 | $ 400 |
| Faculty / Academicians | ₹ 5500 | ₹ 6000 | $ 450 | $ 500 |
| Industry | ₹ 7000 | ₹ 7500 | $ 550 | $ 600 |
| Listeners | ₹ 2000 | ₹ 2500 | $ 300 | $ 300 |

★ Registration fee includes Conference proceedings, Conference Kit, Certificate, Lunch and session breaks.
★ Atleast one author of an accepted paper has to register.
★ Registration fee is to be paid only after the acceptance of the paper. However, Listeners are free to register from now on.
★ Registration form and details of mode of payment are available in the conference weblink **www.karunya.edu/aero/icraae.**
★ Registration fee does not include Journal publishing charges.

### Keynote Address

**Prof. Alon Gany,**
Faculty of Aerospace Engineering,
Technion-Israel Institute of Technology, Israel

### Inaugural Address

**Dr. K. Radhakrishnan,**
Advisor – Indian Space Research Organization
(Former Chairman ISRO & Secretary, Dept of Space)

### Department of Aerospace Engineering

The department of Aerospace Engineering was established in 2009 and is equipped with state-of-the-art facilities such as Subsonic wind tunnel, Supersonic wind tunnel, Kinetic Heating Simulation facility, Shock tube, Water tunnel, CESSNA 152 aircraft etc., The department offers B.Tech and M.Tech in Aerospace Engineering and also P.hD (Full-time and Part-time) with specialization in Aerodynamics, Space dynamics and Heat transfer.

### About Karunya University

Karunya Institute of Technology & Sciences (KITS) was established in the year 1986 by the founders Dr. D. G. S. Dhinakaran and Dr. Paul Dhinakaran. Making rapid progress ever since, the institute has now grown into a university, winning recognitions and awards like the Best Engineering College in Tamilnadu (1996), first ever autonomous self financing Engineering College (2000) and Deemed University (2004) on its way over these years. The institution has excellent academic research and infrastructural facilities sprawl over 700 acres and dedicated faculty, with about 8500 students living on campus. The University offers 13 UG, 20 PG & Ph.D Programmes in almost all the fields of Engineering Technology and Management.

## Chief Patron
**Dr. Paul Dhinakaran**, Chancellor, Karunya University

## Patrons
**Dr. Jeyakumar Daniel**, Chancellor's Representative
**Dr. M. J. Xavier**, Officiating Vice-Chancellor, Karunya University
**Dr. C. Joseph Kennedy**, Registrar, Karunya University

## Chairman
**Dr. G. Prince Arulraj**, Director – Engineering and Technology

## Convenor
**Dr. P. D. Arumairaj**, Head of the Department

## Advisors
**Dr. E. J. James**, Director – International Affairs, Karunya University
**Dr. Pradeep Kumar**, Professor, Aerospace Engineering
**Dr. Ramkrishan Sharma**, Professor, Aerospace Engineering
**Shri. A. N. Subash**, Consultant, Aerospace Engineering

## Organizing Secretaries
**Dr. G. Jims John Wessley**, AP(SG), Aero
**Dr. R. Renjith Singh**, AP, Aero

## Organizing Committee
**Mr. Kalakanda Alfred Sunny**, AP, Aero
**Mr. Aldin Justin Sundarraj**, AP, Aero
**Mr. Prashant Kumar**, AP, Aero
**Lt. S. R. Musica**, AP, Aero
**Mr. M. Gopalsamy**, AP, Aero
**Mrs. S. Parameshwari**, AP, Aero
**Mrs. R. Gayathri**, AP, Aero
**Mr. A. Daniel Antony**, AP, Aero
**Mr. S. Saravanakumar**, Technical Assistant, Aero
**Mr. V. kumar**, Technical Assistant, Aero
**Mr. Y. Selvin Raja**, Administrative Assistant, Aero

## International Advisory / Technical Committee
**Prof. Pini Gurfil**, Technion-Israel Institute of Technology, Israel
**Dr. Prasad. S. Thankabail**, United States Geological Survey, USA
**Dr. Somchai Wongwises**, King Mongkut's University of Technology, Thailand
**Dr. Ahmet Selim Dalkilic**, Yildiz Technical University, Turkey
**Dr. Alberto Coronas**, Universitat Rovira, Spain
**Dr. Suthagar John**, Ibri College of Technology, Sultanate of Oman
**Dr. Raj Das**, University of Auckland, New Zealand
**Dr. I. Dinaharan**, University of Johannesburg, South Africa
**Dr. Lee Chih Fang**, University of Science, Malaysia
**Dr. Pardini Carmen**, Space Flight dynamics Laboratory, Italy
**Dr. Cristina Barrado Muxi**, Polytechnic University of Catalonia, Spain
**Dr. Joseph Gnanaraj**, University of Connecticut, Storrs, USA
**Dr. Rajasekaran**, Jazan University, Saudi Arabia
**Dr. S. J. Vijay**, Jeonju University, Korea
**Dr. Brusly Solomon**, University van Pretoria, South Africa

## National Advisory / Technical Committee
**Dr. M. A. Atmanand**, Chair, IEEE Madras section
**Shri. H. R. Mohan**, Vice Chair, IEEE Madras section
**Dr. A. Nagappan**, Chapter Chair, IEEE Aerospace & Electronic Systems Society
**Dr. G. Rajesh**, Indian Institute of Technology-Madras, Chennai
**Lt. Cdr. John Livingstone**, Johnnette Technologies Pvt Ltd, Noida
**Prof. Anu Jacob Paul**, North Malabar Institute of Technology, Kerala
**Dr. Abhishek**, Indian Institute of Technology, Kanpur
**Dr. V. Ashok**, Vikram Sarabhai Space Centre, ISRO, Trivandrum
**Dr. T. Michael N Kumar**, Bharathiyar College of Engg and Technology, Karaikal
**Dr. Arnalt Stalin**, Bharat Heavy Electricals Limited, Ranipet
**Dr. P. Manjith Singh**, Terminal Ballistics Research Laboratory, Chandigarh
**Dr. Manjunath**, National Aerospace Laboratories, Bangalore
**Prof. A. K. Anil Kumar**, Vikram Sarabhai Space Centre, ISRO, Trivandrum
**Dr. B. C. Pillai**, Professor Emeritus, Karunya University, Coimbatore
**Dr. J. K. Prasad**, Deoghar campus of Birla Institute of Tech, Jharkhand
**Dr. K. Senthil Kumar**, Madras Institute of Technology, Chennai
**Dr. Parag Deshpande**, National Aerospace Laboratories, Bangalore
**Dr. J. Arivudainambi**, Adhiyamaan College of Engineering, Hosur
**Dr. M. Ramji**, Indian Institute of Technology, Hyderabad
**Dr. B. Vinod**, Robotics and Automation, PSG College of Technology, Coimbatore
**Dr. Salih**, Indian Institute of Space Science and Technology, Trivandrum
**Dr. N. Kumarappan**, Secretary, IEEE Madras section
**Dr. P. Sakthivel**, Treasurer, IEEE Madras section
**Dr. K. P. J. Reddy**, Indian Institute of Science, Bangalore
**Shri. Sudhir Kumar Mishra**, Brahmos Aerospace, Bangalore
**Dr. Edwin Raja Dhas**, Noorul Islam University, Nagercoil
**Dr. Jagadeesh**, Indian Institute of Science, Bangalore
**Dr. J. Bruce Ralphin Rose**, Regional Campus, Anna University, Tirunelveli
**Shri. P. Jeevanandam**, Aeronautical Development Agency, Bangalore
**Dr. Sundararaj**, Sri Ramakrishna Engineering College, Coimbatore
**Dr. Noor Danish Ahrar Mundari**, Eternal University, Himachal Pradesh
**Dr. G. Ramesh**, Jain University, Bangalore
**Dr. A. P. Haran**, Park College of Engg and Tech, Coimbatore
**Dr. K. Theerthamalai**, Defence Research and Development Lab, Hyderabad
**Dr. A. K. Verma**, Amity Institute of Space Science and Technology, Noida
**Prof. G. Biswas**, Indian Institute of Technology, Guwahati
**Dr. Rajeev Agarwal**, Sri Aurobindo College, University of Delhi, New Delhi
**Dr. G. Manikandan**, SNS College of Engineering, Coimbatore
**Shri. S. Pandian**, Vikram Sarabhai Space Centre, ISRO, Trivandrum
**Dr. K. Ramachandran**, Bharathiyar University, Coimbatore
**Dr. S. Subha Rani**, PSG College of Technology, Coimbatore
**Dr. G. S. R. Emil Selvan**, Thiagarajar College of Engineering, Madurai

## Accommodation
Accommodation is available at the Karunya Guest House based on the request of the participants. The room tariff is as below :
AC Rooms- Premium : ₹ 2000 (Double occupancy) and ₹ 1500/- (Single Occupancy)
AC Rooms- Deluxe : ₹ 1500 (Double occupancy) and ₹ 1250/- (Single Occupancy)
Non AC rooms : Annexure ₹ 500/-
For accommodation, Kindly contact 0422 2614791.

## Contact
**Dr. G. Jims John Wessley**, Department of Aerospace Engineering, Karunya university.
Email : icraae@karunya.edu | Phone : +91 94878 46614

IEEE Madras Section · BrahMos · इसरो isro · MAFTREE Nurture Engineering Skills

# Peer review statement

All papers published in this volume of *IOP Conference Series: Materials Science and Engineering* have been peer reviewed through processes administered by the proceedings Editors. Reviews were conducted by expert referees to the professional and scientific standards expected of a proceedings journal published by IOP Publishing.

# Application of Anova to Pulse Detonation Engine Dynamic Performance Measurements

Subhash Chander[1], Rakesh Kumar[2], Manmohan Sandhu[3]
TBRL, Chandigarh
India
sckashyapa@gmail.com

TK Jindal[4]
AED, PEC, Chandigarh
India
tkjindal@pec.ac.in

*Abstract*—**Application of Anova to Pulse detonation engine dynamic performance measurement resulted in quantifying engine functionality during various operations. After evaluating the performance, techniques of improving it, were applied in multiple areas of relevant interest. This produced encouraging results and helped in upscaling efforts in a significant manner. The current paper deals in the details of anova implementation and systematic identification of key areas of improvements. The improvements were carried out after careful selection of plans in the engine, ground rig and instrumentation setup etc. It also yielded better reproducibility of performance and optimization of main subsystems of PDE. Further, this will also contribute to reduce the development cycle and trial complexity also, if researchers continue to extract concern areas to be addressed.**

***Keywords— Pulse Detonation Engine, Thrust, Anova, Thrust measurement, Engine ground rig, Defalgration to Detonation (DDT)***

## A. *Abbreviations and Acronyms*

| Abbreviation | Nomenclature | Units |
|---|---|---|
| PDE | Pulse Detonation Engine | - |
| PDRE | Pulse Detonation Rocket Engine | - |
| $P_1$ | Chamber pressure at station 1 | bar |
| PFR | pressure flow regulator | - |
| CFD | Computational Fluid Mechanics | - |
| FCS | fire control system | - |
| ECCU | Electronic command and control unit | - |
| SCR | Silicon - Controlled Rectifier | - |
| AC | Alternating Current | - |
| NI | National Instruments Inc. | - |
| tdms | NI Diadem File storage format | - |
| ANOVA | Analysis of Variance | - |
| $PS_1, PS_2, PS_3, PS_4, \ldots\ldots PS_n$ | Pressures Samples | bar |

## I. Introduction

Pulse detonation engines (PDE) offer a lightweight, low-cost & high thrust alternative for aerospace applications up to hypersonic velocities. PDE is an aerospace engine, which works on detonation principles. It has got few components of simple geometry and have no complex moving parts. Currently, PDE technology is being developed for upper stages that boost satellites to higher orbits. The advanced propulsion technology could also be used for lunar and planetary landers and excursion vehicles that require throttle control for gentle landings. It is also a competitive contender for missile systems application due many other advantages also. It would allow rockets to carry heavier load, reduce costs of launch into orbit too. Even though, PDE is key engine option from last 5 decades for aerospace applications, but a very limited progress has happened in development of fly worthy engine based on these principles.

[1,2,3] – Scientists working in Terminal Ballistics Research Lab, Chandigarh, India
[4] – Associate Professor, Aerospace Engineering Department, PEC University of Technology, Chandigarh, India

In a breakthrough move, Russian scientists have become the first ever to successfully test a pulse-detonation rocket engine (PDRE) in July and August, 2016, on clean fuel (**oxygen-kerosene pairing**), claiming this will be the future of space travel. Americans, Japanese, French and Chinese teams are busy in making PDE powered missions a reality.

TBRL is also engaged in PDE design and is in mature stage to development and testing of ground demonstrator. Getting the desired thrust at all intervals is a key concern of designers and TBRL has put great concentration into this. Detonation achievement phenomenon by using Schelkin spiral blockage ratio optimization was carried out in CFD and experiments to enhance the thrust [1]. During PDE development, designers have faced frequent thrust fluctuations of significant range and emerged as key area to be addressed in addition to integration challenges [2], [3] and [4]. Researchers have earlier developed gaseous PDE ground demonstrator [5] and presently working on liquid fuel based PDE on detonation principles [6]. The power fluctuations will result due random variation of chamber pressure $P_1$. Calibrated and careful selection of various parameters in PDE cycle will help to handle these issues to a large extent. The coming paras is dwelling on our approach to handle this issue too.

## II. Research Methodology - The Analysis of Variance (Anova)

Since, PDE engine is known to have problems of producing consistent thrust without fluctuations, so some techniques are to be applied to identify the phenomenon and address it through carefully drafted strategies [7]. TBRL has applied research methodology of statistical techniques to understand the problem and address it through focused approach [8].

The “analysis of variance” (ANOVA) is used, when comparing three or more groups of numbers [9]. The ANOVA technique is an established practice to understand and isolate the pattern of chamber pressure / thrust variations, so it is explained as under:

### *A. Purpose*

The reason for doing an ANOVA is to see if there is any difference between groups on some variable [9].

ANOVA is available for both parametric (pressure data) and non-parametric (classification/ordering – firing frequency) data.

### *B. Non-parametric and Parametric*

ANOVA is available for score or interval data as parametric ANOVA. This is the type of ANOVA you do from the standard menu options in a statistical package.

The non-parametric version is usually found under the heading "Nonparametric test". It is used when specimen have rank or ordered data.

You cannot use parametric ANOVA, when the data is below interval measurement. Where data is categorical, ANOVA method is not suitable. It would be useful to apply Chi-square technique, which is about interaction rather than about differences between groups.

### *C. Types of ANOVA*

- One-way between groups
- One-way repeated measures
- Two-way between groups
- Two-way repeated measures

TBRL has applied one-way ANOVA to Chamber pressure measured ($P_1$) to analyze the data and provide inferences on the spread of data to reduce the spread of time history of data.

## III. PDE Trial Setup

Experimental PDE trial setup planned by TBRL (refer fig.1) for getting the measurement of various parameters to assess the performance of the PDE [2]. The fuel and air are injected into PDE tube at certain pressure and interval using a unique injection schemes at typical pressures and equivalence ratio established after series of calculations and perfected in experiments. A test rig consisting of purified compressed air cylinder bank, Dome regulator, pressurized flow regulator (PFR) and its allied instrumentation of dial gauges and pressure gauges instrumented through recorders for data storage. This rig also supports to meter and force fuel to be pumped in PDE tube through injector. The current setup helps the PDE input parameter control and regulation.

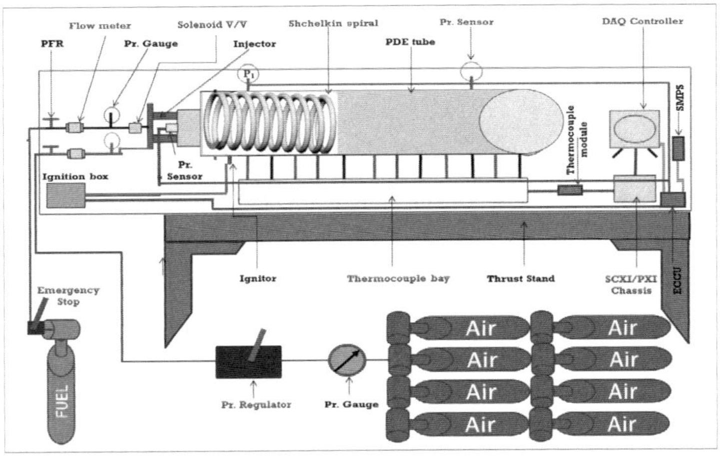


Fig. 1 Trial Setup of PDE experimental plan

The various operations of PDE cycle are carried out in a particular sequence and timing plan, which are altered and fine-tuned to meet our practical requirements. Generally, with a small shift in parameters require retuning of all parameters to get proper output, which is still having performance parameters variations of the order of 20-30% due various reasons. The paper deals with the handling of this issue using complex statistical pattern study and plan to pin point the cause and strategy to address it.

A dedicated fire control system (FCS) is designed to create significant ignition flame spark [5] using SCR based high voltage setup, which is commanded for deflagration / detonation at specific intervals through complex electronic command and control unit (ECCU), inputted remotely through serial link. In present configuration, all these systems are AC powered.

A powerful NI data acquisition system is used to acquire the data from multiple sources at separate sampling rates and being analyzed online and stored in tdms format for detailed offline analysis also. The current paper uses this stored data for doing statistical data analysis.

NI Diadem, Mathworks Matlab & MS Excel is used to carry out data processing, analysis and graphical visualization. This is covered in para V in details.

## IV. Trial Data Analysis using Conventional Statistical Methods

In the current application, peak of tube pressure at stage 1 ($P_1$) in the tube is a key parameter to assess the performance of the PDE. We have used 3 sets of PDE firings at various frequencies are used. These are at 8 Hz, 10 Hz & 12 Hz frequencies. Firstly, 8 Hz frequency data is analyzed and subsequently, the analysis is carried out on all other available frequencies data also.

### *A. Statistical Analysis of 8 Hz $P_1$ data*

The data set used for analysis is 160+ peaks of $P_1$. First to carry out preliminary analysis of data, standard statistical techniques are used. Following assumptions are made to study the pattern:

- Input pressure of fuel and air are uniform throughout the run
- Temperature of tube is constant in the tube
- Ignition pulse spark is of same strength & duration in each shot
- Reflecting shock from the rear end of the Schelkin is not considered
- Pressure sensor sensing element is not in contact with Schelkin
- Effect of vibrations is uniform in all shots

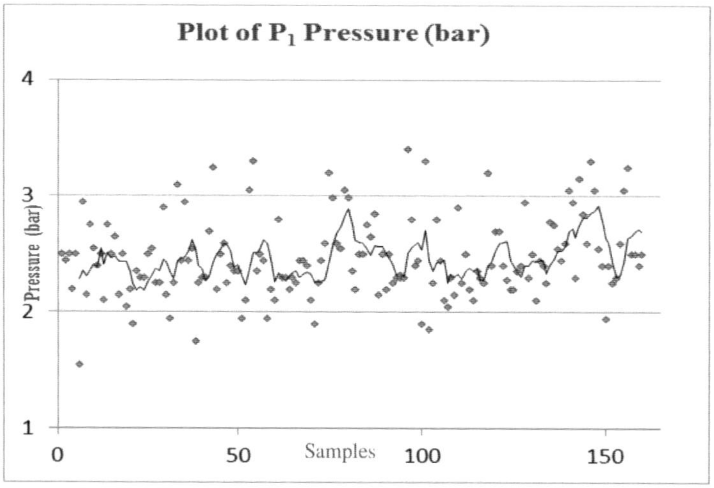


Fig. 2 P1 scatter plot and trend pattern for 8 Hz firing

*1) Sample Spread Analysis*

First, peak values scatter is plotted for 8 Hz firing (figure 2). 4 point moving average based trend line is also inserted for better pattern analysis. The conclusions drawn are shown in table 1.

| Pressure analysis | | | | | Shot analysis | |
|---|---|---|---|---|---|---|
| variation of pr (Bar) | 1.5 to 3.5 (2) | mid value | 2.5 | | missed shots | 0 |
| 20% (>3 bar) | points | 15 | % | 15.5 | higher pr shots | 2 |
| -20% (<2 bar) | | 10 | | | low pr shots | 4 |
| 10% (>2.75 bar) | points | 23 | % | 18.6 | % miss | 0 |
| -10% (<2.25 bar) | | 17 | | | cases consi-dered % | 100 |

Table 1 Pattern analysis of 8 Hz data spread

From the present data analysis, it is established that 18.6 % data is having more than 10% variation spread of pressure. This is one pattern, which is established from current analysis. From trend lines, it is established that trend variation is only from 2.25 to 2.75 (variation band of 0.5 bar only) bar with most of its parameters falling near 2.4 bar.

*2) Average Spread Analysis*

To analyze it further, averaging technique is used [8]. The data is averaged at every one second interval and results are plotted in fig. 3. the formulae used are as per equation, data point, d is

$$d = average\ of\ PS_1, PS_2, PS_3, PS_4, \ldots\ldots PS_n \quad \ldots\ldots (1)$$

Additionally, 4 point moving average based trend line is also inserted for better pattern analysis. Ignoring three peaks, most of the data is very close to 2.4 bar. It means that average values of data follow a significant pattern and is reproducing the repeatability.

From the above analysis, following inferences are drawn:

- The average spread is between 2.3-2.7 bar
- There are 4 insistentent jumps in data
- highest jump is about 0.4 bar
- If ignored, this jump, all other jumps are less than 0.2 bar

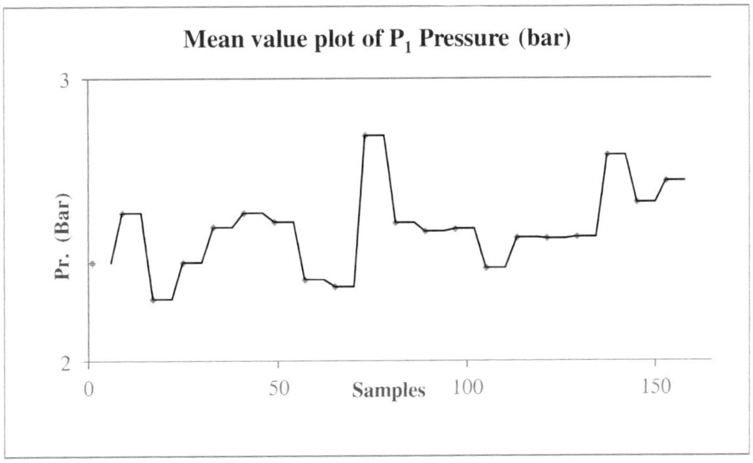


Fig. 3 Avearage P1 plot and trend pattern for 8 Hz firing

From the above analysis, following inferences are drawn:

- The average spread is between -0.3 to 0.3 bar
- There are 3 insistentent jumps in data
- highest jump is about 0.4 bar
- If ignored, this jump, all other jumps are < 0.2 bar
- eventhough data appears to have large spread, but trendline gives more sensible interpetations of the spread of data

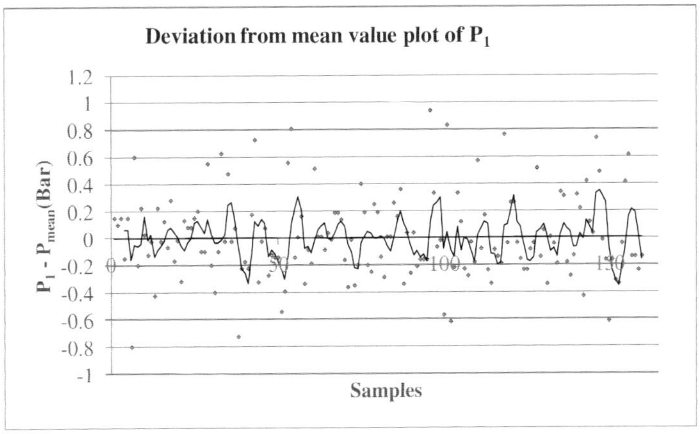


Fig. 4 Devaition of P1 plot and trend pattern for 8 Hz firing

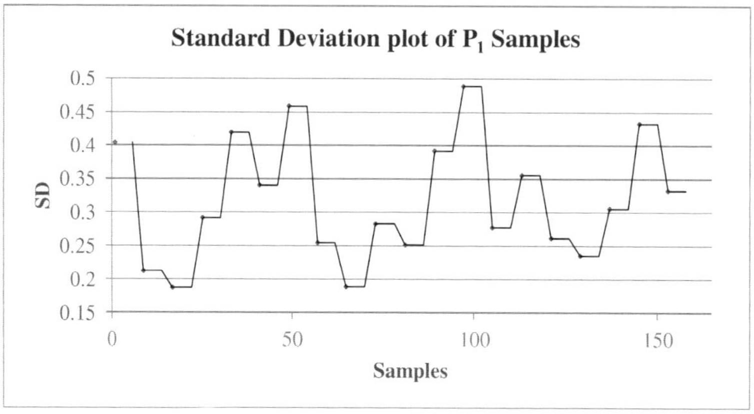


Fig. 5 Standard deviation of error signal ( P1 ) plot and trend pattern for 8 Hz firing

*3) Standard Deviation Spread Analysis*

Standard deviation is established practice to analyze large spread of random data to provide pattern analysis. So, Standard deviation (SD) from mean is calculated from per second mean is calculated and the spread is examined [8]. The SD data spread is plotted in fig. 5. Additionally, 4 point moving average based trend line is also inserted for better pattern analysis of derived data.

From the above analysis, following inferences are drawn:

- The SD spread is between -0.2 to 0.48
- There are 4 insistentent jumps in data
- highest jump is about 0.38
- Eventhough data appears to have large spread, but trendline gives similar interpetations

*4) Histogram Spread Analysis*

Histogram is another method to analyze large spread of random data to provide pattern analysis. So, $P_1$ value histogram is plotted and the spread is examined. The Histogram spread is plotted in fig. 6. It is established that max no. of 46 samples are having values ranging from 2.21-2.43 bar and 44 samples are having values ranging from 2.43-2.63 bar.

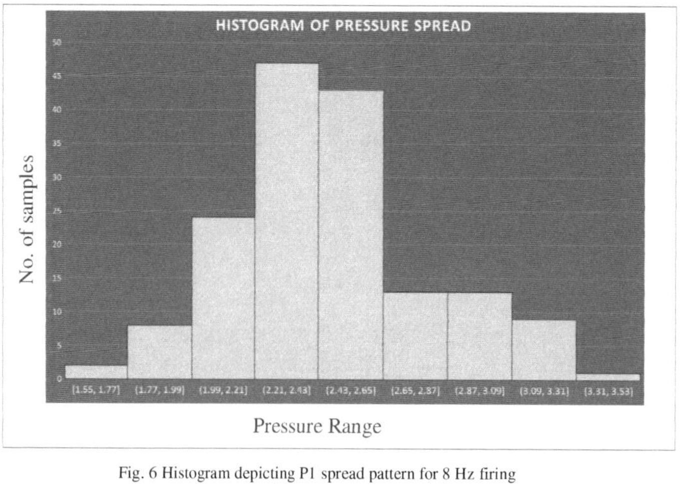


Fig. 6 Histogram depicting P1 spread pattern for 8 Hz firing

*B. Spread Analysis of 8/10/12 Hz $P_1$ data:*

The data set used for analysis is 160+ peaks of $P_1$ deviation from their per second mean. First to carry out preliminary analysis of data, standard statistical techniques of local spread analysis are used. The spread is plotted to highlight the frequency effect in fig. 7. Following observations are made on the data analysis:

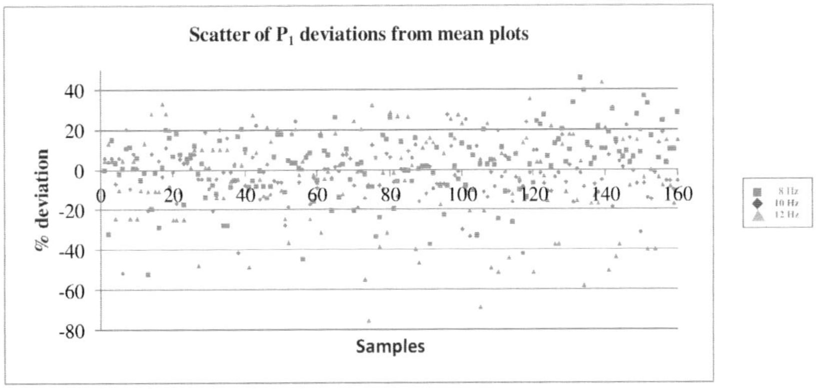


Fig. 7 Scatter plot of 8/10/12 Hz firing

- Spread of data is more significant at higher frequncies
- Missed firing/high pressure shots are more frequent in higher frequencies
- Pressure build-up is also depleted as upscaling of frequencies is planned
- Net impulse is improved with frequncies upscale

Subsequently, the plot % of points falling in various error deviations from per second mean are plotted as under shown in figure 8.

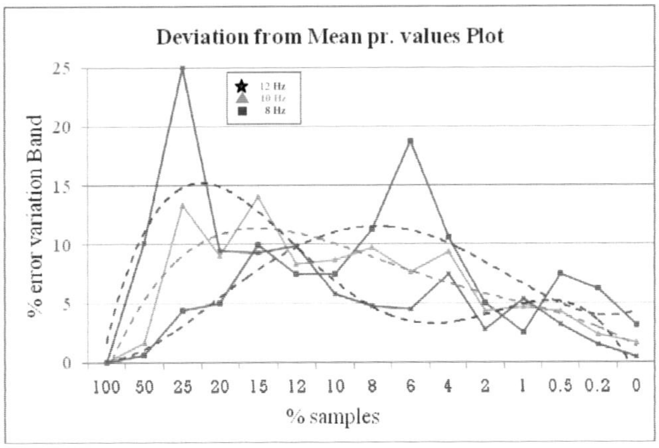


Fig. 8 deviation of $P_1$ from $P_1$ (per second mean) plot with trend pattern for 8Hz / 10 Hz / 12 Hz firing

From this analysis, the interperation are:

1. for 8 Hz, 8 % samples have for than 11 % variation
2. for 10 Hz, 20 % samples have for than 11.5 % variation
3. for 12 Hz, 25 % samples have for than 15 % variation

From this analysis, it is interpreted that as frequency is upscaled, the error magnitude and sample spread falling in higher magnitude is increasing. It is due to the fact that optimization exercises for higher frequencies need to be carried out. Further, the error margins also higher and reasons for this will be dealt with in succeeding paras anlong with anova.

## V. Anova Application To Pde Problem

One-way ANOVA compares three or more unmatched groups, based on the assumption that the populations are Gaussian [9]. In the current problem, 8 Hz, 10 Hz and 12 Hz datasets are three unmatched groups. The assumptions are:

- Each sample is an independent random sample
- The distribution of the response variable follows a normal distribution
- The population variances are equal across responses for the group levels. This can be evaluated by using the following rule of thumb: if the largest sample standard deviation divided by the smallest sample standard deviation is not greater than two, then assume that the population variances are equal.

The key parameters are:

*count*,: ....(2)

....(3)

....(4)

....(5)

| Groups | Count | Sum | Average | Variance |
|---|---|---|---|---|
| 8Hz_dev% | 160 | -234.659 | -1.46662 | 158.2412 |
| 10Hz_Dev% | 299 | 4326.285 | 14.46918 | 189.5913 |
| 12hz_DEV% | 465 | -3905.69 | -8.39934 | 1037.233 |

Table 2 Summary table of 8 Hz / 10 Hz / 12 Hz data

Variability can also be defined in terms of how close the scores in the distribution are to the middle of the distribution. Using the mean as the measure of the middle of the distribution, the variance is defined as the average squared difference of the scores from the mean. Based on the test data and following above assumptions, data is fed to single factor Anova in two steps, and results are tabulated in table 2 for step I (using equation (2) to (5)), subsequently in table 3 for step II as below:

| Source of Variation | SS | df | MS | F | P-value | F crit |
|---|---|---|---|---|---|---|
| Between Groups | 95709.79 | 2 | 47854.89 | 78.29388 | 3.95E-32 | 3.005498 |
| Within Groups | 562934.9 | 921 | 611.2214 | | | |
| Total | 658644.7 | 923 | | | | |

Table 3 Anova table of 8 Hz / 10 Hz / 12 Hz data

### A. *P value*

The P value tests the null hypothesis that data from all groups are drawn from populations with identical means. Therefore, the P value answers this question:

If all the populations really have the same mean (the treatments are ineffective), what is the chance that random sampling would result in means as far apart (or more so) as observed in this experiment.
If the overall P value is large, the data do not give you any reason to conclude that the means differ. Even if the population means were equal, you would not be surprised to find sample means this far apart just by chance. This is not the same as saying that the true means are the same. You just don't have compelling evidence that they differ.

If the overall P value is small, then it is unlikely that the differences you observed are due to random sampling. You can reject the idea that all the populations have identical means. This doesn't mean that every mean differs from every other mean, only that at least one differs from the rest. Look at the results of post tests to identify where the differences are.

In PDE experiments, p value is observed to be 3.95E-32, which is very small, so random sampling differences are observed.

*B. F Ratio and ANOVA table*

The P value is computed from the F ratio which is computed from the ANOVA table.

ANOVA partitions the variability among all the values into one component that is due to variability among group means (due to the treatment) and another component that is due to variability within the groups (also called residual variation). Variability within groups (within the columns) is quantified as the sum of squares of the differences between each value and its group mean. This is the residual sum-of-squares. Variation among groups (due to treatment) is quantified as the sum of the squares of the differences between the group means and the grand mean (the mean of all values in all groups). Adjusted for the size of each group, this becomes the treatment sum-of-squares.
Each sum-of-squares is associated with a certain number of degrees of freedom (df, computed from number of subjects and number of groups), and the mean square (MS) is computed by dividing the sum-of-squares by the appropriate number of degrees of freedom. These can be thought of as variances. The square root of the mean square residual can be thought of as the pooled standard deviation.

The F ratio is the ratio of two mean square values. If the null hypothesis is true, you expect F to have a value close to 1.0 most of the time. A large F ratio means that the variation among group means is more than you'd expect to see by chance. You'll see a large F ratio both when the null hypothesis is wrong (the data are not sampled from populations with the same mean) and when random sampling happened to end up with large values in some groups and small values in others.

The P value is determined from the F ratio and the two values for degrees of freedom shown in the ANOVA table.

F Ratio in PDE experiments is observed to be 78.29388, which shows that value is on higher side, but still in significant region.

*C. Tests for equal variances*

ANOVA is based on the assumption that the data are sampled from populations that all have the same standard deviations. Prism tests this assumption with two tests. It computes the Brown-Forsythe test and also (if every group has at least five values) computes Bartlett's test. Both these tests compute a P value designed to answer this question:

If the populations really have the same standard deviations, what is the chance that you'd randomly select samples whose standard deviations are as different from one another.

The same observations are also observed in PDE experimental data analysis.

More so, $f_{crit}$ is 3.055, which is also stating that samples are wide spread, but however in the acceptable limits.

## VI. Discussion

From the above analysis, it is proved that reproducibility of pressure variations in PDE tube at pressure port 1 is significant, random and repetitive, but within limits. Team has carefully studied the causes of these variations. The main causes are listed below [2]:

- Variation of source line pressures due long pressure lines
- Higher pressure range gauges at rig

- Erratic behaviour of PFR & Dome loaded pressure regulator
- Purity and homogeneity of fuel and air
- Injector issues:
  - design and fabrication issues
  - non homogenous mixing
- Schelkin spiral issues:
  - Mounting mechanism not restrained from vibrations
  - Long length giving shock reflections
  - Random pressure sensor/transmitter touches
- Ignition Command Issues:
  - relay getting stuck or improper contacts
  - Mechanical lags in command following
  - ECCU fire pulse design & time resolution problems
- Ignitor Issues:
  - Body cracking due vibration or ceramic damage
  - Carbon formation at tip
- Misc. Issues
  - heating of tube
  - online variation of pressure
  - contamination
  - sediments in lines and filters
  - leakage and age related problems

All these problems are being addressed simultaneously. The variations in pressures have reduced significantly after addressing key issues of gauges, mounting mechanism of Schlekin spiral, new ignition system design etc. The process of design of injector, ECCU and Fire control system will take one year plus and remaining issues also will get resolved. If these problems are addressed properly, then PDE can be easily to an engine for future aerospace applications [10].

## VII. Conclusions

After this analysis, the team is able to identify the pattern of fluctuation of parameters, which is inherent across globe in PDE technologies. Due which, the focus has moved to correct the various shortcomings in design and trial management to help team to execute time bound assignment of give the system with less than 10 % variation in pressure in 95 % of samples of trial data in operable zone of trial, leaving first one second and last one second. This involves development of new systems and technologies as covered in para VI. By incorporating few of the modifications, the results are encouraging. The system will work within the specified band by improving the identified issues within a span of one year.

## Acknowledgments

The authors are thankful to Director, TBRL for his guidance, encouragement & support to the PDE development. Authors are also thankful to Director, PEC and all members of PDS group of TBRL for their support and participation.

## References

[1] Sanjay Kumar Soni , Amarjit Singh, Manmohan Sandhu, Aashish Goel, Ram Kumar Sharma, **"Numerical simulation to investigate the effect of obstacle on detonation wave propagation in a pulse detonation engine combustor"**, International Journal of Emerging Technology and Advanced Engineering Volume 3, Special Issue 3: ICERTSD-2013, ISSN 2250-2459, pp 458-464, 2013.

[2] Subhash Chander, TK Jindal, **"Integration Challenges in Design and Development of Pulse Detonation Test Rig"** International Journal of Advance Research in Electrical, Electronics and Instrumentation Engineering, Volume 1, Issue 4, ISSN: 2278-8875, DOI : 10.15662/ijareeie, pp 291-304, 2012.

[3] Frank K. Lu, **"Progress and Challenges in the Development of Detonation Engines for Propulsion and Power Production"**, Applied Mechanics and Materials, ISSN: 1662-7482, Vol. 819, pp 3-10, 2016.

[4] Subhash Chander, Tejinder Kumar Jindal, "**Performance Enhancement of Surface to Air Missile by Application of Pulse Detonation Engine based Secondary Propulsion System**", International Journal of Mechanical Engg. (IJME), Recent Science Publications, 2014.

[5] Subhash Chander, Dr. TK Jindal, "**Design of Automated Fire Control System for $C_2H_2/O_2$ Pulse Detonation Rig**", International Journal of Advanced Research in Electrical, Electronics and Instrumentation Engineering, Vol. 2, Issue 2, ISSN: 2278-8875, DOI : 10.15662/ijareeie, pp 924-932, 2013.

[6] Ramamurti, K, **"Explosions and Explosion Safety",** Tata Mcgraw Hills Publishing Co. Ltd., ISBN 9780070704473, 2010

[7] Kothari, CR, Garg, G., **"Research Methodology: Methods and Techniques",** ISBN-13: 978-8122436235, New Age Publishers, 2014.

[8] Montgomery, D.C., Runger, G.C**., "Applied Statistics and Probability for Engineers",** John Wiley & Sons, 2003.

[9] Eva Ostertagová, Oskar Ostertag,**"Methodology and Application of Oneway ANOVA",** American Journal of Mechanical Engineering, Vol. I issue 7, ISSN (Print): 2328-4102, ISSN (Online): 2328-4110, DOI : 10.12691, pp 256—261, 2013.

[10] Subhash Chander, Tejinder Kumar Jindal, "**A study of possible technical aspects needed to be addressed in respect of system engineering of surface to air missile by application of pulse detonation engine**", International Journel of Mechanical Engineering (IJME), Volume 1, Spl. Issue 1, ISSN: 1694-2302, pp 40-43 (2014).

IOP Conf. Series: Materials Science and Engineering **247** (2017) 012002 doi:10.1088/1757-899X/247/1/012002

# PRELIMINARY INVESTIGATION ON THE EFFECTS OF SHOCKWAVES ON WATER SAMPLES USING A PORTABLE SEMI-AUTOMATIC SHOCKTUBE

G. Jims John Wessley
Department of Aerospace Engineering
Karunya University, Coimbatore – 641 114
Jims_john@karunya.edu

## ABSTRACT

The propagation of shock waves through any media results in an instantaneous increase in pressure and temperature behind the shockwave. The scope of utilizing this sudden rise in pressure and temperature in new industrial, biological and commercial areas has been explored and the opportunities are tremendous. This paper presents the design and testing of a portable semi-automatic shock tube on water samples mixed with salt. The preliminary analysis shows encouraging results as the salinity of water samples were reduced up to 5% when bombarded with 250 shocks generated using a pressure ratio of 2. 5. Paper used for normal printing is used as the diaphragm to generate the shocks. The impact of shocks of much higher intensity obtained using different diaphragms will lead to more reduction in the salinity of the sea water, thus leading to production of potable water from saline water, which is the need of the hour.

## INTRODUCTION

A sound wave is a naturally weak wave that causes minor changes in the properties of the fluid along which it travels. The speed of sound is found to be more in solids, quite lesser in liquids and a minimum in gases. The sound waves travel through a gas by the numerous collisions of the molecules in it and these disturbances usually travel with the speed of sound. Moving shock waves are generated when the disturbances due to Collison of molecules is forced through the gas at a speed greater than the speed of sound [1]. Shock tube is a simple device that is used to generate a shock wave in a controlled environment. The driver section in which a gas is compressed to high pressure is separated from the driven section maintained at low pressure or atmospheric pressure by a diaphragm. When the pressure in the driver side exceeds the preset limit, the diaphragm ruptures and causes a shock wave to move along the length of the driven section. These shock waves can be utilized for emerging industrial, commercial, defense, agriculture and home appliance applications like juice extraction, rejuvenating depleted bore wells artificial insemination, treating brain injuries etc.,

## LITERATURE BACKGROUND

For the past few years, scientists have been exploring possible ways to effectively use the shock waves for the betterment of humanity and it is encouraging to see the applications expanding in the recent times. Some of the proven researches with positive outcomes reported in the literature are given below.

Jourdan et.al developed an experimental set-up to investigate the attenuation of a shockwave propagating through a well-characterized cloud of water droplets in a vertical shock tube with different shockwave Mach numbers and droplet diameters. The need for improved quality of visualization and precision of pressure measurements was pointed out in addition to the necessity of improving the physical model [2].

Reddy has developed a simple hand-operated and miniature shock tube, capable of producing shockwaves of Mach 2. The research outlines the design principles and performance capabilities of this mini shock tube, which is more versatile than a traditional compression-driven and blast-driven shock tubes. It is also the first device to produce supersonic flows of sustained duration using human energy, which can generate a Mach more than 1.5. This prototype is a simplified and miniature version of the free-piston driven shock tube (FPST) also known as Reddy's tube which was developed earlier. This miniaturized shock tube has a cylindrical compression tube of length 60 mm length, internal diameter 12.8 mm and a wall thickness of 1 mm [3].

The works of Ulises et.al shows the influence of shock waves on the viability of two strains of bacteria in solution and to verify the impact of shock waves on the microorganisms. It is seen that the lithotripter shock waves do not have significant bacterial effect during shockwave lithotripsy, however bacteria inactivation may be achieved using tandem shock waves [4].

Daosheng Deng et.al performed studies on water Purification by Shock Electro dialysis to filter micron- scale particles and aggregates of nanoparticles present in the feed water which also resulted in the disinfection of 99% of viable bacteria (Escherichia Coli. By combining filtration, separation and disinfection with deionization, it is concluded that Shock Electro dialysis has the potential to demonstrate a highly compact and efficient water purification system [5].

However, it is seen that most of the applications are at the proof-of-concept stage and there has been a consistent effort to build prototypes for suitable commercial applications. Also, the quantum of reported literature on the works carried out is at bare minimum. This paper presents the design and analysis of the performance of a portable shock tube in treating saline water. The outcomes of this work will certainly form a basis for further analysis and enable fellow researchers to optimize the performance and utilize the shockwaves for the betterment of our mankind.

## DESIGN -OF SHOCK TUBE

The main components of the shock tube are a driver section, a driven section and a diaphragm fixed between these sections by means of suitable flanges. The design of a shock tube is based on the initial assumption of the Mach number of the shock to be produced in the shock tube.

The speed of sound is estimated using the relation as below.

$$a = \sqrt{\gamma RT} \tag{1}$$

Velocity of the moving shock at the driven section is estimated as

$$V_s = M_s \times a \tag{2}$$

The pressure ratio and the temperature ratio of the driven and driver section is obtained using the relation as below.

$$V_s = a \times \sqrt{\left(\frac{\gamma+1}{2\times\gamma}\left(\frac{P2}{P1} - 1\right) + 1\right)} \tag{3}$$

and hence

$$\frac{T2}{T1} = \left(\frac{2+(\gamma-1)M^2}{(\gamma+1)M^2}\right) \times \left(1 + \frac{2\times\gamma}{\gamma+1}(M^2 - 1)\right) \tag{4}$$

The diaphragm has to be designed to rupture at any $P_4$, which is estimated as below.

$$\frac{P4}{P1} = \left(\frac{\left(2\gamma M^2 - (\gamma-1)\right)}{\gamma+1}\right) \times \left(1 - \frac{\gamma-1}{\gamma+1} \times \frac{a}{a}\left(M - \left(\frac{1}{M}\right)\right)^{-\frac{2\gamma}{\gamma-1}}\right) \tag{5}$$

The pressure to be maintained in the driver section to obtain shocks of different Mach numbers and their corresponding velocity of the shock with respect to various Mach numbers is shown below in Fig 1.

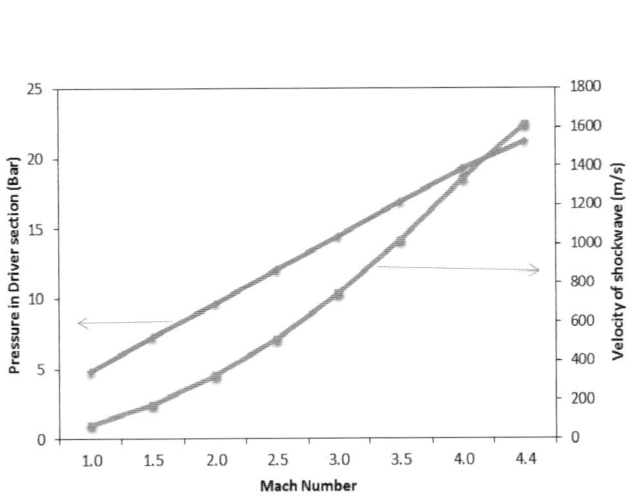


Fig 1. Variation of Pressure and velocity with Mach number

It is seen from Fig 1, that in order to generate a shock of higher Mach number, the pressure required in the driver section increases, while the velocity of the shock also increases linearly. Say for example, to produce a shock of 1.5 Mach, moving with a velocity of 520.9 m/s the pressure needed in the driven section is 2.46 bar. The maximum temperature expected in the driver section due to the maximum pressure in this section which the material should withstand is found from the Fig 2.

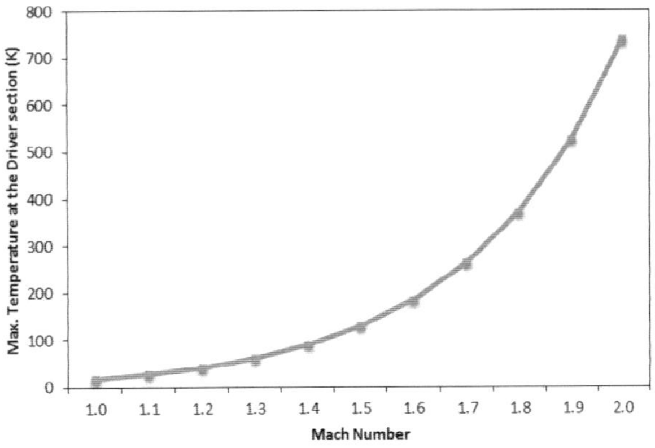


Fig 2. Variation of Pressure and velocity with Mach number

***Material Selection***

Having known the pressure and temperature of operation of the shock tube, it is necessary to choose the right kind of material for its fabrication. The scope of excellent mechanical properties offered by various classifications and grades within the family of stainless steels make them versatile materials for these applications. Stainless Steel SS304 is chosen, as the properties of SS304 suits the requirement in addition to which it also possesses good corrosion resistance, oxidation resistance and it is also easy to form and weld.

The Hoop stress or circumferential stress in the tangential direction of the shock tube is estimated as

$$\sigma_\theta = \frac{Pd}{2t} \tag{6}$$

Based on the hoop stress estimation, the diameter and thickness of the pipe is chosen from the commercially available standard ones. The inner diameter, outer diameter and thickness of the standard pipe for schedule 40 are 20 mm, 26.67 mm and 3.335 mm respectively, which is estimated to withstand a pressure of 29.74 bar. The specifications of shock tube designed are obtained as below in Table 1.

**Table 1. Specifications of the designed shocktube**

| | |
|---|---|
| Driver section Length | 450 mm |
| Driven section Length | 1500 mm |
| Inner Diameter | 20 mm |
| Outer Diameter | 26 27 mm |
| Material | SS304 |
| Diaphragm | Standard A4 paper |

IOP Conf. Series: Materials Science and Engineering **247** (2017) 012002 doi:10.1088/1757-899X/247/1/012002

## EXPERIMENTAL ANALYSIS

The schematic representation of the shock tube facility developed for water treatment is shown in the figure 3.

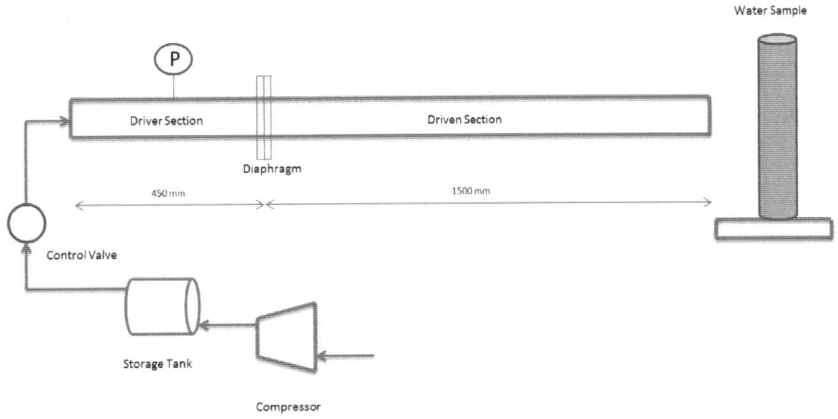


Fig.3 Schematic diagram of the portable shock tube based water treatment facility

The Shock tube receives compressed air from a storage tank where air is compressed and stored. The water samples to be tested are taken in plastic bottles. A control valve is used to adjust the flow of high pressure air flow to the shock tube. A Pneumatic valve is used to open and close the flange that holds the diaphragm. Preliminary investigations were conducted to estimate the bursting pressure of different diaphragms like A4 paper, heavy duty Aluminum foil and Aluminum sheets used for packing foods. The bursting pressures of different diaphragms are found to be as in Table 2.

**Table 2. Estimated bursting pressure for various diaphragms**

| Sl. No. | Diaphragm Material | Thickness of Diaphragm (mm) | No. of diaphragms used | Total thickness of diaphragm (mm) | Bursting Pressure (bar) |
|---|---|---|---|---|---|
| 1 | Standard A4 paper | 0.05 | 2 | 0.1 | 2.5 |
| 2 | Aluminum Foil | 0.016 | 10 | 0.16 | 3.8 |
| 3 | Aluminum packing sheet (Food grade) | 0.2 | 1 | 0.2 | 5.8 |

IOP Conf. Series: Materials Science and Engineering **247** (2017) 012002 doi:10.1088/1757-899X/247/1/012002

The velocity of the shock in the driven section is estimated by recording the peak pressures at two different points separated by a known distance on the driven section by microphones. The output of the microphones is recorded using Sound Forge 6.0 software. Three different water samples were chosen and subjected to shocks and the reduction in salinity is measured using a ppm meter. It is observed that the salt particle starts to deposit on the bottom of the bottle while the pure water starts to accumulate on the top. The outcomes of the preliminary experimental analysis are shown in Table 3.

**Table 3. Reduction of salinity after shock treatment**

| Sample | Type of water | No. of Shocks | TDS Measured Before Shocks (ppm) | TDS Measured after shocks (ppm) | % of TDS reduction |
|---|---|---|---|---|---|
| A | Tap Water (Location 1) | 250 | 582 | 553.2 | 4.9% |
| B | Brackish Water (From Borewell) | 250 | 1908 | 1823 | 4.45% |
| C | Salt Water (20 g/ liter) | 250 | 77300 | 75460 | 2.4% |

## CONCLUSION

A portable shock tube set-up for water treatment is designed and tested with three different types of water samples. It is found that the salinity reduction ranged from 2.4 % to 5% depending on the type of water sample. The salinity reduction in salt water to an extent of 2.4% is a positive indicator that this method can be used to desalinate water by increasing the number of shocks and also the burst pressure. Further investigations are needed to test the quality of water with different diaphragms and higher salinity water samples.

## ACKNOWLEDGEMENT

The authors would like to thank Tamilnadu State Council for Science and Technology for the financial support in completing this project.

**REFERANCES**

1. C. S. Kumar, K. Takayama and K.P. J. Reddy, "Shock waves made simple", Wiley India Pvt. Ltd, New Delhi, 2014.
2. Jourdan.G, Biamino.L, Mariani.C, Blanchot.C, Daniel. E, Massoni.J, Houas.L, Tosello. R and Praguine.D, "Attenuation of shock wave passing through a cloud of water droplets", Shock Waves, Vol 20, pp 285 -296, 2010.
3. K.P.J. Reddy and N. Sharath, "Manually operated piston-driven shock tube, Current Science, Vol 104, No. 2, pp 172-176, 2013.
4. Ulises M. Alvarez, Araceli Ramirez, Francisco Fernandez, Arturo Mendez and Achim M. Loske, "The influence of single-pulse and tandem shock waves on bacteris", Shock Waves, Vol 17, pp 441-447, 2008.
5. Daosheng Deng, Wassim Aouad, William A.Braff, SvenSchlumpberger, Matthew E.Suss and Martin Z.Bazant, "Water purification by shock electrodialysis: Deionization, filtration, separation, and disinfection", Desalination, Vol 357, pp 77-83, 2015.

***************

IOP Conf. Series: Materials Science and Engineering **247** (2017) 012003 doi:10.1088/1757-899X/247/1/012003

# Bandwidth improvement of a rear end slot antenna on a Substrate Integrated Waveguide cavity

Rekha S
Dept of Electronics and communication
Karunya University
Coimbatore, India
anbalaganrekha@gmail.com

M Nesasudha
Dept of Electronics and Communication
Karunya University
Coimbatore, India
nesasudha@karunya.edu

**Abstract** – ***A technique for enhancing the bandwidth of the planar slot antenna on the substrate integrated waveguide cavity has been discussed in this paper. Planar slot antennas are extensively used in research because of its simple profile and moderate gian. Inspite of employing a planar rectangular slot which results in a narrow bandwidth, a broadband slot can be used to improve the bandwidth. This slot antenna is been placed at the rear end of the substrate integrated waveguide and excited by 50Ω grounded coplanar waveguide to provide good isolation. The substrate integrated waveguide adopts the mm wave technology and it facilitates the feeding element and the antenna to be built on the same substrate. The return loss is obtained as -30 dB and -22 dB at 10GHz and impedance bandwidth of 1.98% and 9.9 % for the rectangular slot and the broadband slot respectively. Further the performance is analysed by varying the height of the substrate, the length and width of the slot. The benefits of the antenna include simple profile, low weight, easy fabrication and integration with planar circuits.***

***Keywords — substrate integrated waveguide, slot antenna, broadband, return loss, bandwidth, simple profile and easy fabrication***

## 1. INTRODUCTION

For many decades, slot antennas are highly famous among researchers and industrialists for its simple structure, good isolation from feed network and ease of integration with other planar circuits [1]. They are used for a frequency range of 300MHz to 25 GHz. The slot is a cut on a metal conductor having length in terms of λ. The slot antenna is a complementary of dipole antenna having similar field characteristics except the change in E and H field components. The radiation pattern of a slot is bidirectional. Since slot antenna gives out bidirectional radiation, a metallic reflector or a cavity is placed at a distance of λ/4 on the rear side of the slot to increase the performance in single direction (i.e.) unidirectional pattern. The integration of non-planar metallic cavity at the rear end of the slot makes the system heavier and bulky [2-3].

IOP Conf. Series: Materials Science and Engineering **247** (2017) 012003 doi:10.1088/1757-899X/247/1/012003

The introduction of substrate Integrated waveguide (SIW) helps in integrating planar and non-planar structures over a single substrate with the help of metallic vias which connect the upper and lower plate that forms the side walls of the waveguide. The SIW technology is proposed by cassivi to construct a cavity in a planar substrate [4]. This structure is similar to laminated waveguide [5] or post wall waveguide [6]. Later the SIW cavity has replaced the metallic cavity which is proposed by Luo et al [7]. Researchers have concentrated on size reduction of the metallic cavity by ferrite or dielectric loading in the cavity. In paper [8], with the use of suitable metallic structure the reduction of size has nearly come to half. Further in [9], a Multilayer Printed Circuit Board (PCB) is realized to construct a simple profile cavity backed slot antenna using SIW structure having slot etched at the ground plane and feeding is done through stripline at the middle. The antenna designed on SIW is considered as open ended waveguide radiator [10] and waveguide slotted antenna [11]. Yet all these techniques resulted in a bandwidth of only 2.16% which needs to be improved for real time applications. In [12], a rectangular slot of higher length is used to increase the bandwidth which yields to 6.3%. In this paper, a broadband slot antenna has been designed on SIW to increase the bandwidth and possess moderate gain. The designed slot antenna gives good radiation behavior, simple profile, low cost with ease of fabrication combines the merit of planar slot antennas and cavity backed antennas which can be made through single layer PCB process.

## 2. DESIGN PROCEDURE

The geometry of rear end planar slot antenna is given in Fig 1. It is constructed with four rows of metallic vias arranged linearly. For proper radiation and low leakage, certain rules to be followed as given in eqn (1,2).

$$d/d_2 > 0.5 \quad (1)$$

$$d/\lambda < 0.1 \quad (2)$$

By choosing the conditions [5], the behavior of SIW will be equivalent to metallic cavity. The slot can be etched out at the lower plane to have better isolation from the feeding network. The feeding is given by 50Ω grounded coplanar waveguide to excite SIW. The grounded coplanar waveguide is used as feeding element over the top surface for excitation of the slot on the ground plane. The top and bottom plate is interconnected by metallic vias which forms the structure of substrate integrated waveguide.

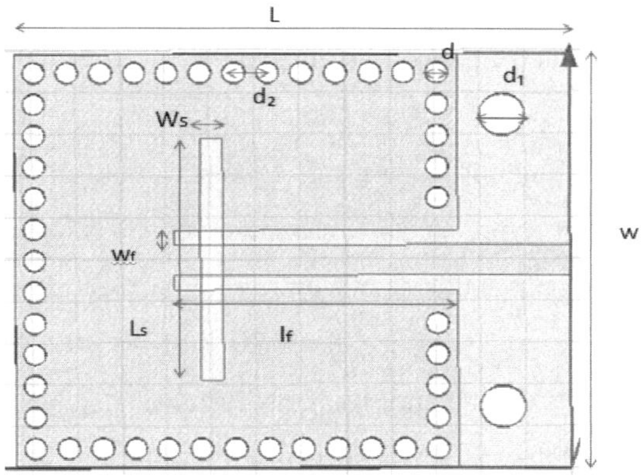


*Fig. 1. Geometry of rear end rectangular planar slot antenna on SIW*

The geometric parameters of the antenna are referred from the paper [7] as given in Table 1. The return loss of the antenna is given in fig 2 as -30 dB at 10 GHz and it is a narrow band antenna. This shows that the antenna can be suitable for an application where more isolation is needed for other frequencies except 10GHz.

Table 1 Geometrical parameters of the antenna in fig 1

| **PARAMETER** | **NOTATION** | **UNIT (in mm)** |
|---|---|---|
| Size | $l \times h \times w$ | 19.8 x 23.8 x 0.5 |
| Diameter of vias | $d$ | 1 |
| Diameter of big vias | $d_1$ | 2 |
| Centre to centre of vias | $d_2$ | 1.5 |
| Width of slot | $w_S$ | 1 |
| Length of slot | $L_S$ | 11.2 |
| Length of feed | $L_f$ | 11.5 |
| Width of feed | $W_f$ | 0.9 |

To increase the bandwidth of the slot antenna, the planar rectangular structure is replaced by a broadband slot as shown in fig 2.

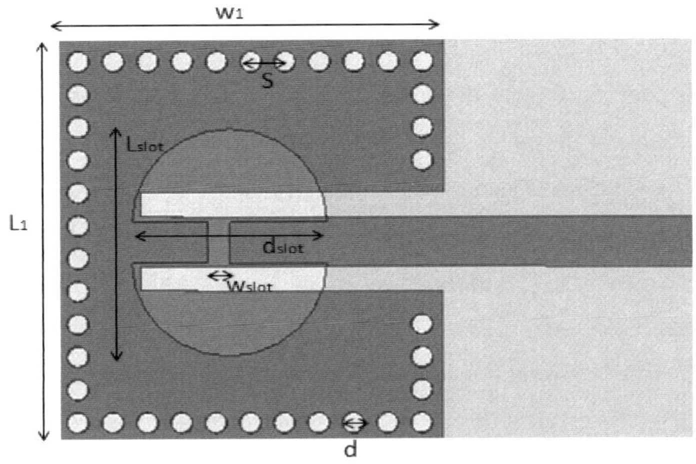


*Fig. 2. Geometry of rear end broadband planar slot antenna on SIW*

The broadband slot antenna is etched on the bottom layer of SIW structure and the 50 ohm grounded co-planar waveguide is etched out on the top layer. The SIW is formed by rows of metallic vias inserted into a dielectric filled rectangular waveguide. To avoid leakage losses between metallised vias eqn (1) and (2) has to be maintained. The geometry of fig 2 is given in Table 2.

Table 2 Geometrical parameters of the antenna in fig 2

| **PARAMETER** | **NOTATION** | **UNIT (in mm)** |
|---|---|---|
| Size | $l$ x $h$ x $w$ | 19.8 x 30 x 0.8 |
| Diameter of vias | $d$ | 1 |
| Centre to centre of vias | $S$ | 1.6 |
| Diameter of slot | $d_{slot}$ | 8 |
| Width of slot | $w_{Slot}$ | 1 |
| Length of slot | $L_{Slot}$ | 12 |

## 3. ANTENNA PERFORMANCE

(a) Performance of planar rectangular slot antenna:

The antennas has been simulated on Rogers/RTDuroid 5880 substrate having the permittivity of 2.2 and loss tangent 0.001 and copper is chosen as the metal on the upper and the lower plate. The substrate height is less than λ/50 which makes the size of the cavity smaller than the conventional cavity. The metallic vias is also formed by copper. It is simulated for continuous frequency of 7 to 12 GHz and return loss is been found to be -30 dB at 10GHz as given in fig 3. The antenna is simulated by varying the height of the substrate and length of the slot, the return loss characteristics is been noted which is shown in fig 4 and fig 5. It is shown that as the height of substrate increases, the bandwidth also increases slightly. It is clear that the cavity depth is directly related to impedance bandwidth from Fig 4. The bandwidth is nearly measured to be 200 MHz. The slot has higher effect on the resonant frequency. By increasing the slot length, the resonant frequency is decreased to lower region as shown in Fig 5. It is inferred from the result that resonant frequency is inversely proportional to the slot length. It is good to keep slot length as half of wavelength for better radiation efficiency. It is possible to vary width of the slot to improve impedance bandwidth. As the width of the slot increases, the bandwidth increases to a little extent and when the width is decreased the bandwidth is decreased and hence the width is directly proportional to the bandwidth as shown in fig 6.

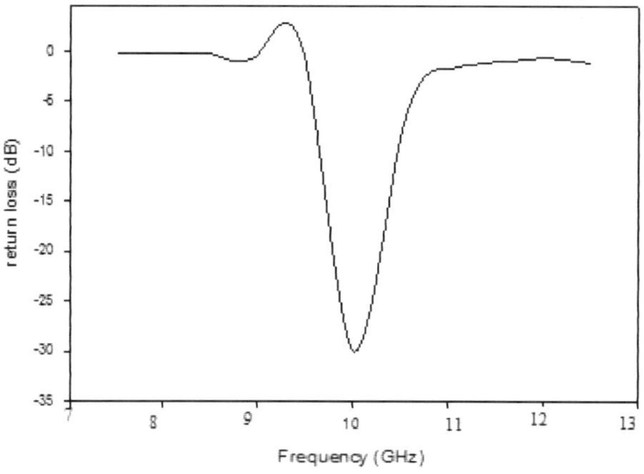

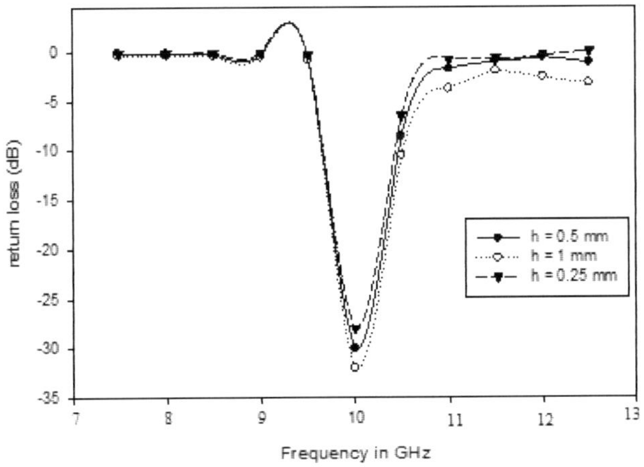


*Fig. 4. Return loss of the rectangular slot antenna by varying height of the substrate*

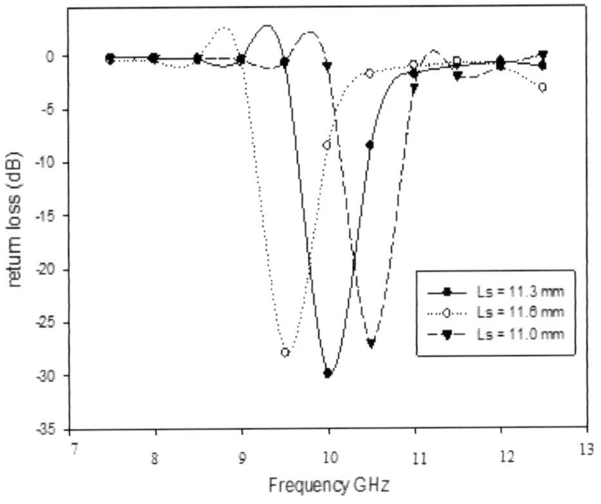


*Fig. 5. Return loss of the rectangular slot antenna by varying length of the slot*

IOP Conf. Series: Materials Science and Engineering **247** (2017) 012003 doi:10.1088/1757-899X/247/1/012003

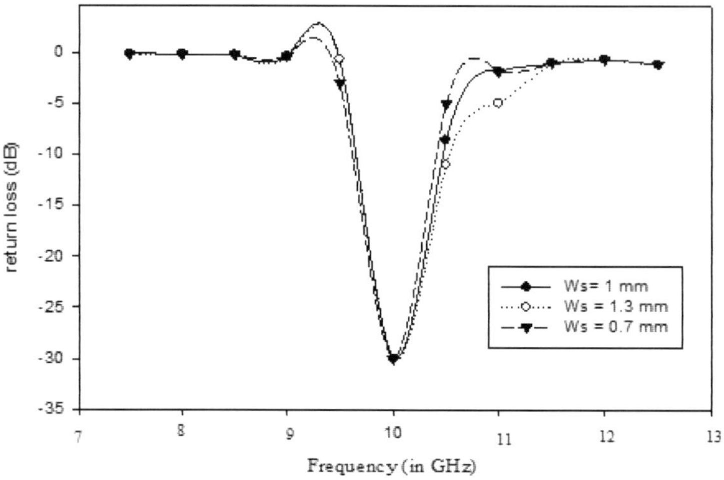


*Fig. 6. Return loss of the rectangular slot antenna by varying width of the slot*

(b) Performance of planar broadband slot antenna:

Rogers/RTDuroid 5880 substrate having permittivity of 2.2 and loss tangent of 0.001 is been used to simulate the broadband slot antenna. The top and bottom conductor is chosen to be copper. The height of the substrate is chosen to be 0.7mm to retain the simple profile configuration. The return loss is simulated to be -22 dB at 10 GHz and the 10 dB impedance bandwidth is found to be 1GHz (9.9%) from 10.5 to 9.5 GHz which is better than the bandwidth of rectangular slot antenna as shown in fig 7.

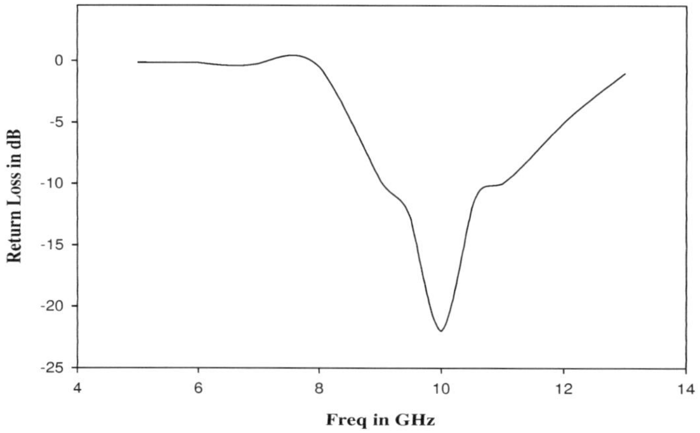


*Fig. 7. Return loss of the broadband slot antenna*

IOP Conf. Series: Materials Science and Engineering **247** (2017) 012003 doi:10.1088/1757-899X/247/1/012003

The bandwidth and the input impedance of the broadband slot antenna firmly depend on the diameter of the slot ($d_{slot}$). The effect of $d_{slot}$ on the operating bandwidth is shown in fig 8 in which the 10 dB impedance bandwidth raises with increase in $d_{slot}$ until the impedance matching of feeding network is not disturbed. The resonant frequency is affected by the variation in the length of the slot ($L_{slot}$). As the $L_{slot}$ increases, the resonant frequency shifts towards the lower end of the frequency as both the parameters are inversely proportional. So the resonant frequency is determined by the length of the slot as shown in fig 9.

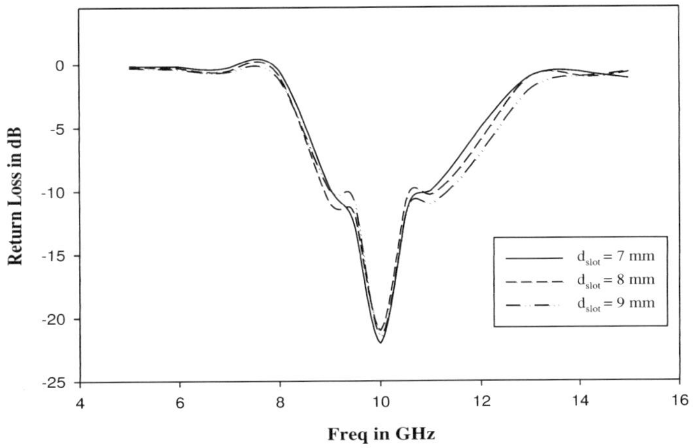


*Fig . 8. Return loss of the broadband slot antenna by varying diameter of the slot*

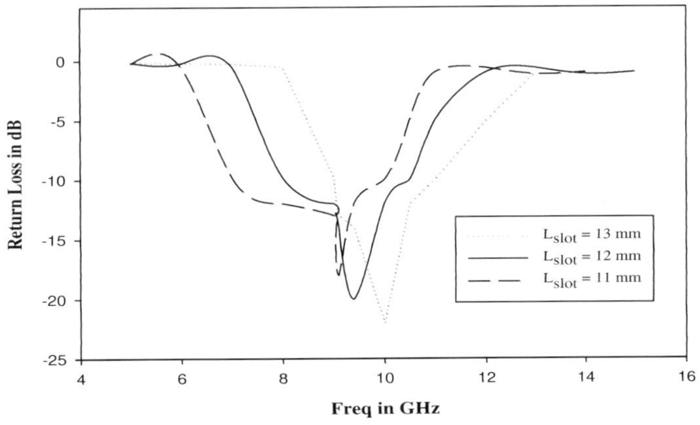


*Fig . 9. Return loss of the broadband slot antenna by varying length of the slot*

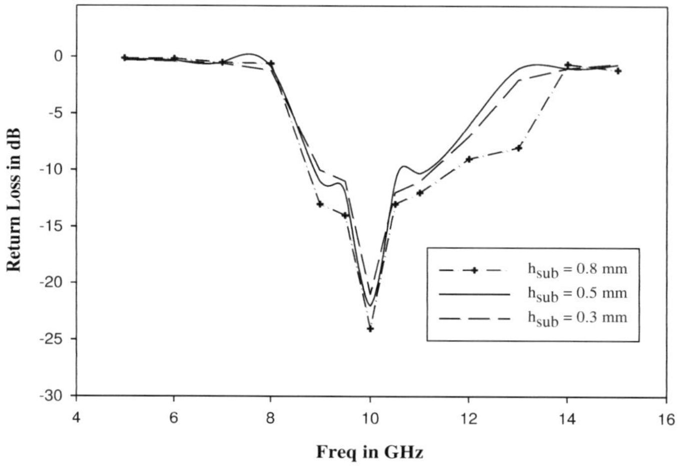


*Fig . 10. Return loss of the broadband slot antenna by varying height of the substrate*

It is shown in fig 10 that as the height of the substrate is reduced the bandwidth diminishes to an extent and when the height of substrate is increased the bandwidth increases unless there is no change in impedance matching. The antenna can be designed according to the specifications by altering the physical parameters.

(c ) Comparison of performance:

The antenna1 and 2 are made of identical substrate, metallic plates, feeding element and other electrical characteristics. A major contrast is the shape of slot antenna as rectangular slot and broadband slot. The rectangular slot gives a bandwidth of 200 MHz whereas the broadband slot gives a enhanced bandwidth of 1GHz because of its broadband structure, The effect of variation in height of the substrate, length of the slot and diameter (width) of the slot in both the antenna is non-controversial to each other. The gain of the broadband slot antenna is found to be 3 to 3.5 dBi in the operating range whereas the gain of the rectangular slot is 4.6 to 5 dBi because of the narrow radiating property of rectangular slot. The fabrication of SIW slot antenna can be done by Printed Circuit Board (PCB) technology.

## 4. CONCLUSION

A rear ended planar rectangular and broadband slot antenna fed by grounded coplanar waveguide has been designed on a single substrate technology namely SIW. The designed slot is low profile, simple and effective when compared with conventional cavity back slot antenna. The return loss is -30 dB and -22 dB at 10GHz for rectangular slot and broadband slot respectively. The bandwidth in the broadband slot is found to be 1GHz (9.9%) whereas it is 200MHz (1.98%) for rectangular slot. The performance variation is analyzed in accordance with change in height of the cavity, length and width of the slot. It is studied from the results that return loss and impedance bandwidth can be improved by the variations in height of substrate than the changes in width of slot. Likewise the length of the slot has indirect proportionality with the resonant frequency of the slot antenna. The slot antenna can be varied parametrically in accordance with the specification of the design and application.

REFERENCES

[1] Yoshimura, Y.: 'A microstripline slot antenna (short papers)', *IEEE Transactions on Microw. Theory and Techniques*, 1972, *20*, (11), pp. 760-762

[2] A. Adams, "Flush mounted rectangular cavity slot antennas-Theory and design," in *IEEE Transactions on Antennas and Propagation*, 1967, 15, (3), pp. 342-351

[3] Vallecchi, A., and Gentili, G. B.: 'Microstrip-fed slot antennas backed by a very thin cavity', *Microw. Opt. Technol. Lett.*, 2007, 49, pp. 247–250

[4] Cassivi, Y., Perregrini, L., Wu, K.: 'Low-cost and high-Q millimeter-wave resonator using substrate integrated waveguide technique' Proc. European Microwave Conf., 32, 23 September 2002

[5] Uchimura, H., Takenoshita, T., Fujii, M.: 'Development of a laminated waveguide', *IEEE Trans. on Microw. Theory Techn.*, 1998, 46, (12), pp. 2438-2443

[6] Hirokawa, J., Ando, M.: 'Single-layer feed waveguide consisting of posts for plane TEM wave excitation in parallel plates', *IEEE Trans. Antennas Propag.*, 1998, 46, (5), pp. 625-630

[7] Luo, G. Q., Hu, Z. F., Dong, L. X., et al.: 'Planar slot antenna backed by substrate integrated waveguide cavity', *IEEE Antennas Wirel. Propag. Lett.*, 2008, 7, pp. 236-239

[8] Hong, W., Behdad, N., Sarabandi, K.: 'Size reduction of cavity-backed slot antennas', *IEEE Trans. on Antennas and propag.*, 2006, *54,* (5), pp. 1461-1466

[9] Locker, C., Vaupel, T., Eibert, T. F.: 'Radiation efficient unidirectional low-profile slot antenna elements for X-band application', *IEEE Trans. on Antennas and propag.*, 2005, *53,* (8), pp. 2765-2768

[10] Clenet, M., Litzenberger, J., Lee, D.: 'Laminated waveguide as radiating element for array applications', *IEEE Trans. on Antennas and propag.,* 2006, *54,* (5), pp. 1481-1487

[11] Yan, L., Hong, W., Hua, G.: 'Simulation and experiment on SIW slot array antennas', *IEEE Microw. Wirel. Compon. Lett.*, 2004, 14, (9), pp. 446-448.

[12] Luo, G. Q., Hu, Z. F., Li, W. J., *et al.*: 'Bandwidth-Enhanced Low-Profile Cavity-Backed Slot Antenna by Using Hybrid SIW Cavity Modes', *IEEE Trans. on Antennas and Propag.*, 2012, 60, (4), pp. 1698-1704

IOP Conf. Series: Materials Science and Engineering **247** (2017) 012004 doi:10.1088/1757-899X/247/1/012004

# Progression In The Concepts Of Cognitive Sense Wireless Networks – An Analysis Report

**Ajay.V.P** (*Author*)
Department of Electronics and Communication Engineering
Kumaraguru College of Technology
Coimbatore, India
ajay.vp.ece@kct.ac.in

**Nesa Sudha.M** (*Author*)
Department of Electrical Sciences
Karunya University
Coimbatore, India
nesasudha@karunya.edu

*Abstract* – **This paper illustrates the conception of networks, their primary goals (from day one to the present), the changes it had to endure to get to its present form and the developments which are in progress and in store for further standardization. The analysis gives more importance to the specifics of the Cognitive Radio Networks, which makes use of the dynamic spectrum access procedures, framed for better utilization of our available spectrum resources. The main conceptual difficulties and current research trends are also discussed in terms of real time implementation.**

***Keywords — wireless networks, cognitive radio, spectrum management, spectrum access, dynamic decision, game theory, wireless security, internet of things***

## I. INTRODUCTION

Fixed spectrum analysis and allocation of usage bands went together in governing the usage methods of the available spectrum ranges for so long that it was thought to be the pinnacle of wired and wireless networking concepts. However, with the advent of more effective network equipment manufacturing procedures and the increase of potential users for them, it was later understood that the pinnacle was not at its highest point with the fixed policies. Newer strategies and better variations in the spectrum assignments were needed in order to compensate for the usage losses. With this in mind, researchers were able to bring about many new changes and one such successful paradigm is the Dynamic Spectrum Access (DSA) conditions for spectrum allocations. Sporadic and drone-like utilization of spectrum decreases the efficiency and performance of the network, even to as high a percentage as 85%.

Dynamic Spectrum Access prohibits situations like the ones shown in Figure 1 and Figure 2, by opportunistic methods. [1] [2] The field of wireless communication and networks have gained so much importance that almost all existing technologies are extensively used in all possible scenarios to find the best technology suited for providing cognitive capabilities in the latest 4G, 5G and LTE based systems. Researchers are further comparing different mathematical and analytical models to bring out the best decision and detection model for Cognitive Radios over the past decade. To carry out all these factors in real time, data concerned with the current spectrum usage and their license information needs to be thoroughly analysed and modified for present scenarios and conditions. Most channels, being fading channels, need special decision models and data logging in order to estimate the channel effectiveness under overlay methods (Primary user or Secondary user access to spectrum) and underlay conditions (Primary and Secondary users together with special considerations).

This needs a collection of prerequisite information on when the spectrum is free, what type of priority is needed for each part of the base network, the type of users present and methods of managing the network effectively. This paper shows the past and present functionalities perfected for the effective network architecture, by providing adequate details on spectrum mobility, sensing and management for sharing mechanisms. An overview of the DSA and CRNs, along with an investigative report on the influence of all such functionalities is also being presented here. As a final touch, the design possibilities are also discussed with suggestions to improve the present status of CRNs [2].

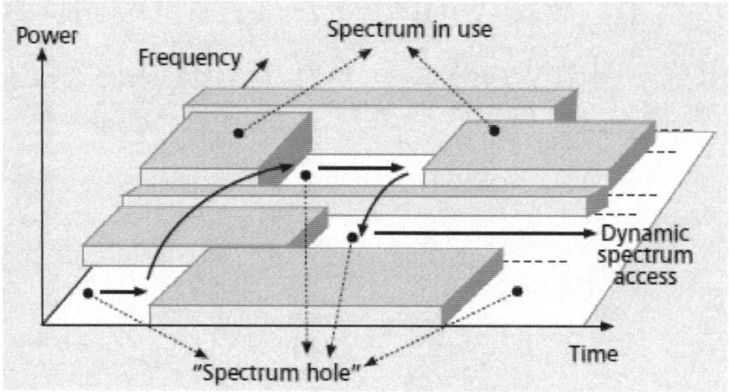


*Figure 1* – Spectrum-hole concepts

## II. EVOLUTION OF WIRELESS NETWORKS TO ITS PRESENT FORM

Though the present generation of networks craves the usage of wireless networks for simplicity and easier implementation, they are to be used in careful ways so as to avoid wastage and loss of resources when there are already so little to spare. The users who first made use of wireless networks were uncomfortable with the amount of spectrum utilized for even one single transaction of information to be successful. This lead to doubts whether all could get resources from the communication spectrum in the days to come. After all, 10 kHz to 30GHz cannot be sufficient for all kinds of communication techniques which are currently in use and for those which are being brought into the market every other day.

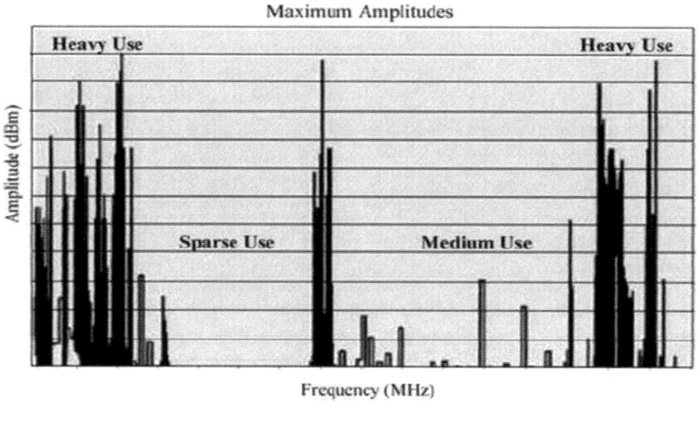


*Figure 2* – Spectrum under-utilization

Thus, people turned their attention to the "use and reuse" concept for their wireless networking resources. The first step towards this concept is the advent of Frequency reuse principle, which made the sharing of spectrum frequencies possible between geographically separate plains. It helped users to share spectrum resources between predefined distances and still be efficient in providing the required QoS for the users' requirements.

But this still was not enough for feeding the enormous appetite of the wireless users and better solutions were required from service providers. Methods like dynamic frequency allocations, sharing of frequencies and dynamic licensing of spectrum resources and so on are popularly being implemented these days. In the midst of such innovative designs and theoretical endeavors, the most commonly studied concept is that of the Cognitive Radio Networking and Technology, which is the heart of this research study paper.

## III. COGNITIVE RADIO TECHNOLOGY

Dynamic accessing of spectrum is being heavily studied for the past three years and many articles, research findings and theoretical ideas are being deployed for the scientific community's usage. One of the methods used for fine-tuning the working of the DSA (Dynamic Spectrum Access) is the Cognitive Radio Networks. [3]

The power to provide the capability to opportunistically, share the licensed wireless channel with others is the key to the working concept of DSA. Effective spectrum management techniques are needed if high bandwidth is to be provided for the mobile users' around the required regions, who are all under different versions of wireless (heterogeneous) architectures.[

To control the unavoidable fluctuations in the spectrum allocations between users, the following characteristics are given to the CR networks;

- Ability to find out free portions of an available spectrum
- Ability to choose the best channel amongst them
- Ability to coordinate the channel access between users
- Ability to vacate the occupied channel when it is needed by the primary user

Thus, these abilities can be coined under the following techniques meant for CR system;

- Spectrum sensing
- Spectrum decision
- Spectrum sharing
- Spectrum mobility

Also the system needs to be implemented in such a way that no major modifications are required to be done from the already existing systems, to avoid increase of expenses.

With these points in mind, this paper is aimed at providing simplified information about the ways of spectrum management and the techniques incorporated in the CR system as mentioned above.

Two main characteristics may be shown with regard to this definition;

- *Cognitive Sense* – Identification of unused and idle spectrum is done with real time information collected in real time from the environment. The best channel spectrum is then selected from them and utilized in such a way that there is no interference to the licensed user in way possible
- *Multi-Configurations* – Reconfigurability of transceiver to use different ways of access methods that are supported through the proper hardware design.

A CR needs to use an optimized transceiver architecture system with two main components – a radio front section and a baseband processing unit. The Front section deals with the amplification, mixing and A/D conversion for the received signal, while the Baseband processing section is necessary for modulation and demodulation of the converted signal. The real time environment changes can be dealt with, by providing a control bus for reconfiguring the components as and when needed by the time-varying conditions.

Another important requirement of a CR technique is the presence of a wideband RF front section which will be useful for sensing simultaneously a wide range of frequencies. Factors that govern the working of the front section would be the RF hardware design, power amplifier, wideband antenna and so on. The system should thus have the ability to detect faint signals as well.

The main factors behind a CR technique may be well explained with the help of its definition;

***"Radio which can change its transceiver parameters with respect to the environment's interaction and the predefined priorities of the users"***

## IV. COGNITIVE RADIO NETWORKS

A complete description for the CR network architecture is first needed, before entering the facts behind the same. The CR network has two basic modules – a primary (or) licensed network and secondary (or) CR network.

The licensed network is the existing network, which has primary users operating at a certain predefined frequency spectrum. Control for the primary network is given through base stations, which weeds out the presence of unauthorized users in the network.

The secondary or CR network is a network which does not have the required license to operate at a particular spectrum band.

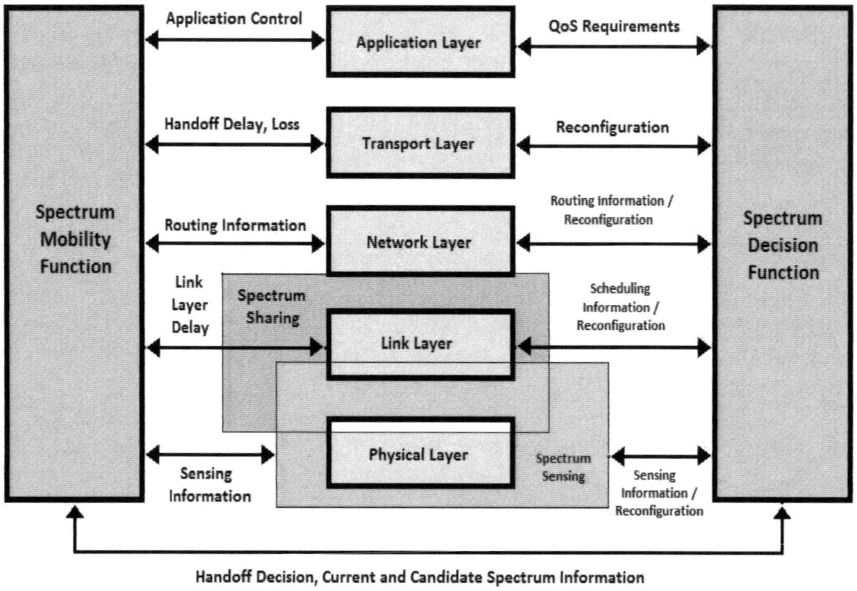


*Figure 3* – Spectrum management for Cognitive Radio

They are allowed opportunistically, to access the network resources through separate single hop CR base stations.[7] Spectrum brokers are spectrum resources distribution agents included in the CR side of the network. Spectrum utilization by the CR networks may be shown to be both in the licensed regions of the primary user spectrum and in the unlicensed regions through wide access band technologies. Such network operations may be divided into two main types like licensed and unlicensed operations.

The licensed operation is done by the primary user network, focusing mainly on the ways to detect primary users. This is to help the system allocate and vacate the resources as and when needed. The unlicensed operations are responsible for providing the same privileges as that of the primary users when they are not in the system, through some spectrum sharing mechanism.

CR users utilize three different types of access methods;

- CR Network access – CR network users are allowed to access the specific CR base station on all spectrum bands
- Adhoc CR access – Communications from one CR user to the other is done through adhoc connection in both the spectrum band types
- Network primary access – Accessing the PR base station by the CR users, by using an adaptive MAC protocol

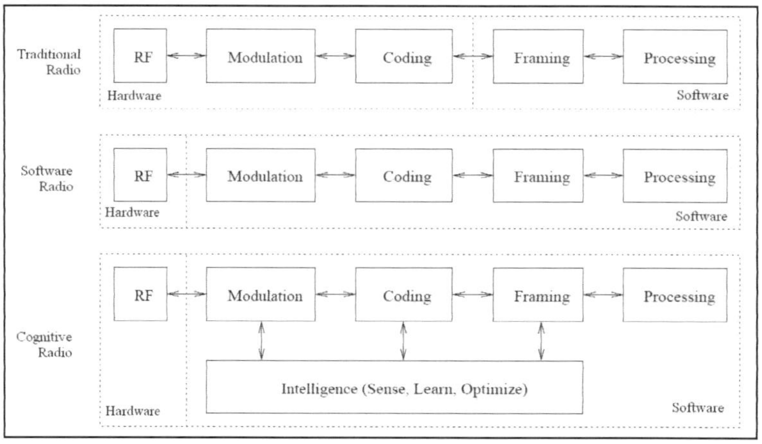


*Figure 4* – Comparison for Traditional Radio, Software Radio and Cognitive Radio

## V. STRUCTURE FOR SPECTRUM MANAGEMENT

Due to the unique nature of the CR based networks, spectrum management conditions are needed to face all the design challenges, such as,

- Avoiding interference – Interference between primary and CR user networks need to be kept zero
- Updated QoS knowledge – The CR networks need to support communication between them for transfer of information regarding the QoS parameters for the heterogeneous and dynamic environment
- Smooth communication – Communication between the CR networks need to be smooth, even during the presence of primary network users

For implementing such challenging grounds in the CR system, a framework for the spectrum management [8] is needed. The management requires four major functions, in general, for all kinds of CR network architectures;

- Sensing – Determining unused / idle frequency portions for the sake of a CR user, through monitoring and detecting of spectrum holes
- Decision – Allocation of channel to CR users through the implementation of predefined policies
- Sharing – Spectrum sharing between CR users without collision between them
- Mobility – Ability to continue the CR user's work in another vacant spectrum region, when the current spectrum region is needed by the licensed user

In cognitive radio system, unlicensed secondary users can use the resources which are licensed for primary users. When primary users want to use their licensed resources, secondary users has to vacant these resources. Hence secondary users have to constantly sense the channel for detecting the presence of primary user. It is very challenging to sense the activity of spatially distributed primary users in wireless channel. Spatially distributed nodes can improve the channel sensing reliability by sharing the information and reduce the probability of false alarming. [6]

Similar to the importance of the architecture management issues, there are many sub-functions meant for spectrum management, under each function;

### *A. Sensing*

One of the most important functions needed for a CR network environment is the presence of a sensing algorithm, which is designed solely for informing the CR about the changes taking place in the entire system.

With sensing capabilities, the system will be able to adapt itself and detect spectrum holes effectively, thereby eliminating primary network interference.

The sensing techniques that are used here will be of three categories; [4]

#### *a. Detection of primary user transmission*

CR user observations are responsible for the detection of transmission and such detections are done by,

- Matched filter based detection – By incorporating the primary user characteristics into the matched filter, it will serve as a detector for the primary user signals
- Detection of energy – In the event of insufficient information from the primary user about its signal characteristics, it is best to use detection methods based on the presence of energy patterns. But the method is prone to false alarms and uncertainty of correct results
- Detection of inherent features – Periodic flow of signals is detected and analysed first; a special correlation spectral function is used for the same. The method is robust and certain of better results, but poses to be a very complex method to implement.

Some of the major problems involved here are,

- Lack of interactions between PR and CR users
- Interferences (external, internal and environmental)
- Line of sight and shadowing problems

Hence, it is best to go in for the method of cooperative detection in such cases, to get a more accurate detection of primary user transmission. It is both advantageous and inefficient at the same time – it is best suited due to its robust nature, its ability to avoid shadowing and multipath fading, while it is not capable of reducing the network overhead for such a heavy traffic. [5]

*b. Detection of primary user reception*

Even if the cooperative detection pattern is useful in reducing interference, the most common method for detecting spectrum holes is to detect the users in the primary network within the range of a CR user, by using the leakage power of the receiver system. However, it is only possible in TV receivers due to the fact that weak local oscillator leak is not enough for other forms of receiver detection.

*c. Management of Interference temperature*

Control measures for interference in the transmitter side can be governed with the help of analyzing the power radiated and the position of transmitters. But since, interference happens at the receivers, a separate model called interference temperature is considered and the same is used to bring down the interference by the setting a limit for it.

Now the CR users are allowed to use the band until the Interference temperature does not reach beyond the specified limit.

However, fixing a plausible limit for the interference temperature is not easy.

Challenges involved in sensing;

1. Measurement of Interference temperature – cannot be fixed easily
2. Sensing for networks with multiple users – difficult to implement
3. Sensing through spectrum efficiency – decreases efficiency as CR users need to stop sensing to allow sensing

*B. Decision*

To equip a CR network to decide the best channel for the user, the system needs to be incorporated with the ability to consider QoS requirements based decision making. This is termed as spectrum decision. This term, however, has little importance in research papers, due to the fact that most people consider some rule for minimal allocation strategies without putting efforts into this part. Difficulty in decision making arises due to the presence of multiple CR users in the system, interference from other users, faulty access and sensing methods used and finally due to the complex format needed to implement such a procedure.

It is essential to know more about the characteristics of the channel, procedures for decision making and challenges involved in decision making.

*a. CR networks' channel characteristics*

All spectrum holes present in the system should be considered for providing a time-varying environment along with parameters related to the spectrum. Such parameters are as follows;

- Interference – Useful for estimating the capacity of the channel and for deriving the power of a CR user
- Path loss – Depends on the frequency involved and the distance. Path loss increases with increase in the frequency of operation, which in turn decreases the range of transmission.
- Errors in the wireless links – Changes with each modulation technique used and with interference levels too
- Delay via link layer – When the system uses different types of link layer protocols, delays are incurred. So identification of spectrum bands for combining the parameter characterizations together.

*b. Procedures for decision making*

Decision making involves reconfiguring of the mode of transmission and the bandwidth needed for such a transmission. There is no guarantee for a spectrum band to be available for a CR user during the entire period of time, as primary user has more priority in the system. In this context, the CR users will not be able to use a particular channel indefinitely and as this is true for all CR network cases, decision making needs to be dynamic, as well as continuous in nature.

Noncontiguous spectrum bands are used for the transmission in CR networks simultaneously. Thus, this method can help form signals with high throughput and which is not deterred by primary users and interferences. So when one of the bands is undergoing handoff, the rest of system remains in the same condition.

*c. Challenges involved in decision making*

As in the cases of sensing, there are a few common challenges in the process of decision making;

- Model for decisions – Since SNR based spectral estimation is not enough, different QoS requirements pose a difficulty for the framing of a compatible model of decision making procedures
- Reconfigurable cooperation – Reconfiguration, being an integral part of the system of decision making, helps in changing parameters as and when needed. So reconfiguration is to be combined with the cooperation among users is difficult.
- Heterogeneous spectrum decision making – Making a CR network to accommodate both the licensed and unlicensed bands alike is not easily done

*C. Sharing*

Sharing of wireless channels in CR networks needs the cooperation of transmission attempts among the users, which shows the importance of MAC protocol in sharing. Some of the most high spoken for challenges in the sharing procedures are coexistence of users, range of spectrum available and the actual user priorities involved.

The four main aspects of the sharing procedures in CR networks may be explained as;

1. Sharing procedure architectures
2. Spectrum allocation ways
3. Techniques for spectrum access
4. Intra/Inter networking

*a. Sharing procedure architectures*

The main architecture models in sharing will be either decentralized or centralized; centralized sharing mechanism will be controlled only with the help of one control entity while the decentralized mechanisms will have many node based local or global policies rather than a single version of standards.

*b. Spectrum allocation ways*

Two ways of allocating shared spectrum will be cooperative type and non-cooperative type;

Cooperative sharing of spectrum will make sure that the interference from each node is considered, in order to form a cluster based network model for balanced sharing.

Distributed or non-cooperative sharing methods are selfish solutions for the CR network systems, as they do not communicate between one another. Performance is degraded due to non-cooperative approaches while cooperative techniques prove to be effective.

*c. Techniques for spectrum access*

Other forms of sharing techniques include overlay and underlay spectrum sharing mechanisms; overlay method deals with the sharing of spectrum with CR users only in absence of primary users in the spectrum, while in underlay mechanisms the CR users are allowed to use the spectrum even during the presence of primary users, with the spread spectrum techniques. Such techniques will make sure that the presence of CR users will be regarded as noise by the primary users.

*d. Intra / Inter networking*

- Intranetworking mode of sharing – Focuses on the sharing and allocation of spectrum between primary and CR users alike. The CR users are permitted to try accessing the channel, even in the presence of primary user, without causing interference.
- Internetworking mode of sharing – Overlapping system locations enabled via multiple CR systems, governed by policies.

*e. Challenges involved in sharing*

Some of the common spectrum sharing challenges involved in the tuning of a CR network may be given as follows;

- Channel based common control – A CCC or a common channel control is needed in sharing of spectrum but it is not possible to use a fixed CCC in real time systems, for single or clustered systems
- Range of the actual dynamic radio – As the radio operators and their frequencies are not dependent, the consequent changes in the frequency of operation is possible, which cannot be addressed always
- Basic unit of spectrum - Without considering the spectrum as single unit, the sharing is not possible in real time and this consideration is not easily included in our algorithms.

- Information on the location of the CR user – An assumption that the CR users know the location of primary users and vice versa cannot always be proved right when it comes to real time systems

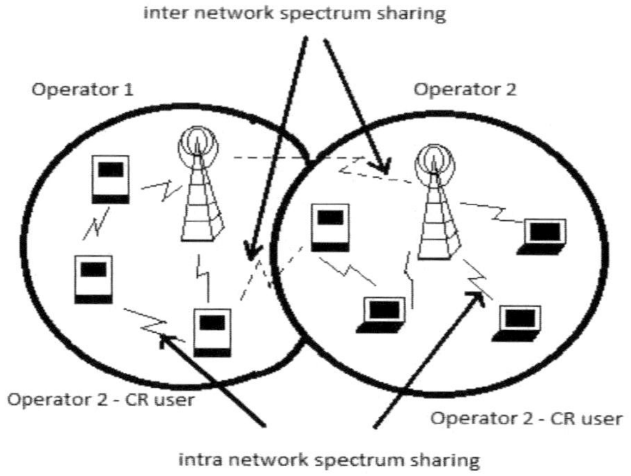


*Figure 5* – Intra / Inter networking

*D. Mobility*

Managing the CR network users' change of operating bands, when the primary user requires that band, is an ultimate requirement of the concepts of Cognitive Radio networks. It does not mean that once a CR user locates a perfect band in the operating spectrum, it can be used by the same user indefinitely or even till the end of the transmission.

The primary user will sometimes immediately require the use of a particular spectrum band which was allocated by the decision algorithm to the CR user. This update of priority should make sure that the CR frequency is released and handed over, back to the licensed owner.

Mobility management is the key to the handoff procedures, which we use currently in many situations. Mobility management ensures that the transitions are smooth and quick, without degrading the QoS of the primary user. Since this procedure requires the attention of most layers in the system, it is also in need of using actual real time information related to the duration of handoff procedures. So spectrum mobility and spectrum handoff are still open for research ideas and improved implementation / investigation.

*a. Challenges involved in mobility*

Two most important problems seen in terms of mobility are,

- Time domain related mobility – Enabling of QoS in time varying channel availability is difficult
- Space domain related mobility – Enabling of QoS when user moves from one place to another is also not easily possible

VI. CURRENT RESEARCH TRENDS IN COGITIVE RADIO NETWORKS

The notion of providing decision making intelligence to already existing technologies and systems to achieve Cognitive Radio Capabilities has given rise to several prospects of research in this field. Some commonly referred terminologies in this field of research are Emergency system, Cooperative spectrum sharing, non-cooperative spectrum auction, spectrum efficiency and estimation, Dynamic decision models and Spectrum safeguards. Some of the highly noted research interests in this domain (but not limited to) may be discussed as follows;

*a. Evaluating Cognitive Radios*

Cognitive radios are being deployed and implemented in most parts of scientific community with which many claim to have solved the spectrum access problems. However, the authenticity of such claims and the physical proof of the same needs to be verified as pointed out in [6], [10] and [20].

*b. Dynamic Spectrum auction and access*

Auctions for spectrum access, under the governing bodies of spectrum utilization in each region of the world, are carried out for both primary access as well as cognitive access. The hierarchical structure of such access are framed during that time and provided to respective takers of the portion of spectrum.

However, the auctioned spectrum needs to be verified from time to time for potential misuse of the policies. Some papers like [11] and [22] gives a detailed account on how auctions for spectrum access are done, how policies for their usages are framed and the steps necessary for controlled strategies to use the shared spectrum effectively. [15] shows the opportunistic behavior and outage statistics of the spectrum usage model with bidirectional relay users control.

*c. Emergency Services and Crowd control*

[21] shows the collection and manipulation of data concerned with calamity struck regions in order to guide people away from the hazardous location quickly. Random neural networks, Dynamic spectrum utilization and reinforcement learning are used for calculating the best guided routes for different categories of the people with different parametric variations. Similar usage of Cognitive Radio models in the field of Emergency services are researched upon for real time implementation and commercialization of such Cognitive radio modules.

*d. Emulation of Cognitive Radios with single / multiple primary and secondary users*

Number of primary users in the spectrum corridor plays a major role in determining how well the spectrum can be shared with secondary users and to what extent the secondary users can effectively use the shared spectrum without sacrificing quality of service on both ends of the system. The emulation of Cognitive radio with importance to the number of primary users is shown in [16] and to some extent in [11].

Similarly uncontrolled number of secondary users in the system contributes to induced interference to the primary users and also to self-interference. The statistics in such study reports show that the number of primary users in comparison with the number of secondary users can be a very optimal area of interest for research in this domain.

*e. Future Network Throughput enhanced models*

In real time Radio scenarios, it is almost impossible to determine the entropy of the system and the prediction of users is therefore uncertain most of the time. [19] shows such environments which exhibit imperfect conditions for user prediction and resource allocation in the system. [31] provide vital results pertaining to the context of imperfect sensing ad the throughput calculations involved in those conditions. The results show the effect of multiple users and Hybrid Automatic repeat requests on the Cognitive Radio model.

[27] states the importance of cloud based data logging and control with improved throughputs for the test conditions involved. It also shows that researchers are combing the concepts of Internet of Things with the Cognitive Radio models to achieve connected emergency services.

*f. Security aspects and behavioral pattern logging*

Due to the fact that receiver signal is random and fading in nature, the transmission power is fixed high enough at the source which can be utilized by eavesdroppers to determine the data sent or behave like an authorized primary user in the system by studying the signal parameters carefully. Such conditions are discussed in [18] with emphasize on MISO systems.

Attacks to the resources of the Cognitive radio environment takes place in all lower layer levels of the OSI model and the most common attack region is the physical layer. Spoofing attacks [40], Denial of Services and Data Falsification attacks [42] are some recently studied attack models which are seriously undermining the research in this field. Detailed listings and countermeasures of many such attacks have been discussed and elaborated in [41] and [42].

*g. Spectrum access to cloud services and QoS*

Cloud based architectures for opportunistic uploads and downloads [33] and Quality of services measurement in relay based machine level commuications [34] are some of the latest trends in Cognitive Radio research domains. Many research aspects have been identified in the area of Quality of services enhancement and evaluation for Cognitive radio network. The main idea is to find out the effectiveness of a proposed system in order to make commercialization of the same a lot easier. Also papers like [41] show that the Quality of services is severely affected by user attacks on the system.

*h. Power parameter measurements for different CR models and user numbers*

Energy analysis, Power controls and thresholds settings for primary and secondary user detection are shown to be alternative options of research in recent years. [17] shows spectral utilization for certain scenarios in real time and calculates estimation of parameters for modeling into a real time installable system. Power usages for specific routing algorithms [26] and transmission strategies for Cognitive wireless sensors [35] are also researched upon by many. Such results provide useful references to model a good decision model and dependable version of Radio networks, especially for 802.22 Emergency services.

*i. Optimal Spectrum Sensing, Mobility and Allocation management*

Research in the four major functional domains of Cognitive Radio working model will be another viable option at all times, as newer technologies and systems require different and better tools to operate. The four main functions as discussed in the previous chapter of this survey are Sensing[12][32], Decision[14][36], Sharing/ Management[16][37][38] and Mobility. Researchers study the aspects either an individual function or in combination with one or more of the functions together. Also Queuing concepts and mutual cooperation among users with reference the four functions are also being analyzed deeply.

*j. Transmission and Reception model for MISO and MIMO Cognitive Radios*

Newer concepts like Multiple Input and Single / Multiple Outputs are being analyzed for usage along with Cognitive radio networks as they tend to improve QoS and Transceiver capabilities of a wireless channel. The signal conditioning, multi-copy transmission and SNR adjustments can be done during transmission from source which are discussed and elaborated in [24] and [25]. [36] accounts for the same type of calculations with better considerations in the form of separate sensor nodes and imperfect sensing probabilities.

While transmission parameters play an important role, the parameters concerned with the receiving end have an equally important role towards MISO and MIMO Cognitive Radios. [18], [23] and [29] show us the cooperation of Cognitive systems in terms of MISO, MIMO and Massive MIMO.

*k. Interference management algorithms in cooperative models*

The detection of spectrum underutilization primarily uses three mechanisms namely, energy detection, cyclostationary methods and matched filtering. Out of these three main methods, energy detection is used commonly for the secondary user access in the absence of primary user in case of Overlay access methods or along with primary users in the case of Underlay access methods. The interference between primary user and secondary user as well as within different secondary users can be a vital key to finding whether the spectrum is in use or not. [15], [30] and [38] show the above points in detail with particular emphasis on primary users' interferences.

*l. Resource utilization statistics for existing commercial Cognitive Radio hardware platforms*

The most difficult type of research for CR domain concepts, in general terms, will be to choose commercially available modules and to test the same under various algorithms and design aspects by either fixed parametric calculations or through analytic and experimental scenarios. There are hardware platforms for Cognitive Radio implementations like USRP modules from Ettus research which can be of great use for real time analysis of our algorithmic models.

Companies like Nokia [45], xGTechnologies [44], Alcatel, Ettus Research [49], Wipro [43], Xilinx [48] and many others have been in the field of Cognitive Radio Development for quite some time now and each of their platforms have their own advantages and disadvantages.

Software tools specifically designed for cognitive radios like GNU Radio [46] [47], SDR and libraries written for CR implementations that can be interfaced with existing software platforms are also available in the market.

## VII. CONCLUSION

Opportunistically exploiting the spectrum for CR network needs helps achieve total quality of service and to utilize the limited spectrum, while reducing inefficiency in the usage criteria. With the capabilities of a CR network, a system with a complete knowledge of the spectrum is created. In this survey, the main goals of CR networks along with most of the sub category protocols and functions were explained. This survey also deals with the complete set of explanations for all related functions of a CR network such as sensing, decision making, sharing and mobility management. Several factors are playing key roles in the research methodologies of Cognitive radio in real time implementation and some of the common aspects were discussed in brief. Hence, in order to provide an effective communication system with spectrum aware properties, better research results and surveys are needed.

## REFERENCES

[1] http://www.dyspan-sc.org/
[2] ET Docket No 03-322 – *FCC Notice of Proposed rules making and ordering* – Dec 2003
[3] Simon Haykins, "*Cognitive Radio: Brain Empowered wireless communications*" IEEE JSAC volume 23. No.2, Feb 2005
[4] D.Cabric, S.M.Mishra and R.W.Brodersen, "*Implementation Issues in Spectrum Sensing for Cognitive Radios*" Proc. 38th Asilomar Conf. Sig., Sys. and Comp. 2004, Nov 2004

[5] I.F.Akyildiz et al., "*NeXt Generation / Dynamic Spectrum Access / Cognitive Radio Wireless Networks: A survey*" Comp. Networks J., vol 50. Sept 2006
[6] S. M. Mishra, A. Sahai, and R. W. Brodersen, "*Cooperative Sensing among Cognitive Radios*," Proc. IEEE ICC 2006, Vol.4, June 2006
[7] J.Ansari, X.Zhang and P.Mähönen, "*A Decentralized MAC for Opportunistic Spectrum Access in Cognitive Wireless Networks*" ACM 2010
[8] H. Zheng and L. Cao, "*Device-centric Spectrum Management*," Proc. IEEE DySPAN 2005, Nov. 2005
[9] Yao Xie, Armbruster, B and Yinyu Ye, "*Dynamic Spectrum Management With the Competitive Market Model*" Trans. on Signal processing, April 2010
[10] Mahfuzulhoq Chowdhury, Asaduzzaman, and Md. Fazlul Kader, "*Cognitive Radio MAC Protocols: A Survey, Research Issues and Challenges*", Smart Computing Review, Vol.5, No.1, February 2015
[11] Jikai Yin and Qi Shi, Li Li, "*Bid Strategies of Primary Users in Double Spectrum Auctions*", Tsinghua Science and Technology, ISSN 1007-0214 04/12 pp152-160, Volume 17, Number 2, April 2012
[12] Nan Zhao, Fangling Pu, Xin Xu and Nengcheng Chen, "*Optimisation of multi-channel cooperative sensing in cognitive radio networks*", Published in IET Communications, 24th May 2013 (doi: 10.1049/iet-com.2012.0748)
[13] Marko Kosunen, Vesa Turunen, Kari Kokkinen and Jussi Ryynänen, "*Survey and Analysis of Cyclostationary Signal Detector Implementations on FPGA*", IEEE Journal On Emerging And Selected Topics In Circuits And Systems, Vol.3, No.4, December 2013
[14] Muhammad Junaid Farooq, Muddassar Hussain, Junaid Qadir and Adeel Baig, "*A Game-Theoretic Spectrum Allocation Framework for Mixed Unicast and Broadcast Traffic Profile in Cognitive Radio Networks*", 38th Annual IEEE Conference on Local Computer Networks, Citigate Central Sydney, Australia, 21 Oct - 24 Oct 2013
[15] Xing Zhang, Zhenhai Zhang, Jia Xing, Rong Yu, Pengwei Zhang, and Wenbo Wang, "*Exact Outage Analysis in Cognitive Two-Way Relay Networks With Opportunistic Relay Selection Under Primary User's Interference*", IEEE Transactions On Vehicular Technology, Vol.64, No.6, June 2015
[16] Panagiotis-Agis Oikonomou-Filandras, Kai-Kit Wong, and Yangyang Zhang, "*Informed Scheduling by Stochastic Residual Belief Propagation in Distributed Wireless Networks*", IEEE Wireless Communications Letters, Vol.4, No.1, February 2015
[17] Fourat Haider, Cheng-Xiang Wang, Harald Haas, Erol Hepsaydir, Xiaohu Ge and Dongfeng Yuan, "*Spectral and Energy Efficiency Analysis for Cognitive Radio Networks*", IEEE Transactions on Wireless Communications, Vol.14, No.6, June 2015
[18] Jun Xiong, Dongtang Ma, Kai-Kit Wong and Jibo Wei, "*Robust Masked Beamforming for MISO Cognitive Radio Networks With Unknown Eavesdroppers*", IEEE Transactions On Vehicular Technology, Vol.65, No.2, February 2016
[19] Jian Yang and Hangsheng Zhao, "*Enhanced Throughput of Cognitive Radio Networks by Imperfect Spectrum Prediction*", IEEE Communications Letters, Vol.19, No.10, October 2015
[20] Jared J. Thompson, Kenneth M. Hopkinson and Mark D. Silvius, "*A Test Methodology for Evaluating Cognitive Radio Systems*", IEEE Transactions On Wireless Communications, Vol.14, No.11, November 2015
[21] Olumide j. Akinwande, Huibo Bi and Erol Gelenbe, "*Managing Crowds in Hazards With Dynamic Grouping*", IEEE Access, Vol.3, July 24, 2015 (DOI 10.1109/ACCESS.2015.2453341)
[22] Angeliki V. Kordali and Panayotis G. Cottis, "*A Contract-Based Spectrum Trading Scheme for Cognitive Radio Networks Enabling Hybrid Access*", IEEE Access, Vol.3, September 10, 2015 (DOI 10.1109/ACCESS.2015.2455492)
[23] Lifeng Wang, Hien Quoc Ngo, Maged Elkashlan, Trung Q. Duong and Kai-Kit Wong, "*Massive MIMO in Spectrum Sharing Networks: Achievable Rate and Power Efficiency*", IEEE Systems Journal, accepted June 1, 2015
[24] Zhijin Qin, Yue Gao and Clive G. Parini, "*Data-Assisted Low Complexity Compressive Spectrum Sensing on Real-Time Signals Under Sub-Nyquist Rate*", IEEE Transactions On Wireless Communications, Vol. 15, No. 2, February 2016
[25] Sami Akin and Markus Fidler, "*On the Transmission Rate Strategies in Cognitive Radios*", IEEE Transactions on Wireless Communications, Vol. 15, NO. 3, March 2016
[26] Ahmed Al-Saadi, Rossitza Setchi, Yulia Hicks and Stuart M. Allen, "*Routing Protocol for Heterogeneous Wireless Mesh Networks*", IEEE Transactions on Vehicular Technology, Vol. 65, No. 12, December 2016
[27] Qihui Wu, Guoru Ding, Zhiyong Du, Youming Sun, Minho Jo and Athanasios V.Vasilakos, "*A Cloud-Based Architecture for the Internet of Spectrum Devices Over Future Wireless Networks*", IEEE Access, June 24, 2016 (Digital Object Identifier 10.1109/Access.2016.2576286)
[28] Yuan Ma, Yue Gao, Ying-Chang Liang and Shuguang Cui, "*Reliable and Efficient Sub-Nyquist Wideband Spectrum Sensing in Cooperative Cognitive Radio Networks*", IEEE Journal On Selected Areas In Communications, Vol. 34, No. 10, October 2016
[29] Manman Cui, Bin-jie Hu, Xiaohuan Li, Hongbin Chen, Shiwei Hu and Yide Wang, "*Energy-Efficient Power Control Algorithms in Massive MIMO Cognitive Radio Networks*", IEEE Access, (DOI 10.1109/ACCESS.2017.2652441)
[30] Koya Sato and Takeo Fujii, "*Kriging-based Interference Power Constraint: Integrated Design of the Radio Environment Map and Transmission Power*", IEEE Access / IEEE Transactions On Cognitive Communications and Networking, 2016 (DOI. 10.1109/ TCCN.2017.2653189)
[31] Ateeq Ur Rehman, Lie-Liang Yang and Lajos Hanzo, "*Delay and Throughput Analysis of Cognitive Go-Back-N HARQ in the Face of Imperfect Sensing*", IEEE Access (DOI 10.1109/ ACCESS.2017.2649380)
[32] António Furtado, Luís Irio, Rodolfo Oliveira, Luís Bernardo and Rui Dinis, "*Spectrum Sensing Performance in Cognitive Radio Networks With Multiple Primary Users*", IEEE Transactions On Vehicular Technology, Vol. 65, No. 3, March 2016
[33] Yegui Cai, F. Richard Yu and Shengrong Bu, "*Dynamic Operations of Cloud Radio Access Networks (C-RAN) for Mobile Cloud Computing Systems*", IEEE Transactions On Vehicular Technology, Vol. 65, No. 3, March 2016

[34] Shih-Chun Lin and Kwang-Cheng Chen, "*Cognitive and Opportunistic Relay for QoS Guarantees in Machine-to-Machine Communications*", IEEE Transactions On Mobile Computing, Vol. 15, No. 3, March 2016
[35] Wensheng Zhang, Cheng-Xiang Wang, Di Chen and Hailiang Xiong, "*Energy–Spectral Efficiency Tradeoff in Cognitive Radio Networks*", IEEE Transactions on Vehicular Technology, Vol. 65, No. 4, April 2016
[36] Shao-Chou Hung, Yong Xiao and Kwang-Cheng Chen, "*Transmission Strategy With Cooperative Sensors in Cognitive Radio Networks*", IEEE Transactions On Vehicular Technology, Vol. 65, No. 5, May 2016
[37] Xiang-Yang Li, Panlong Yang and Yubo Yan, "*SPA: Almost Optimal Accessing of Nonstochastic Channels in Cognitive Radio Networks*", IEEE Transactions On Mobile Computing, Vol. 15, No. 6, June 2016
[38] Ali Y. Al-Zahrani and F. Richard Yu, "*An Energy-Efficient Resource Allocation and Interference Management Scheme in Green Heterogeneous Networks Using Game Theory*", IEEE Transactions On Vehicular Technology, Vol. 65, No. 7, July 2016
[39] Nam-Phong Nguyen, Chinmoy Kundu, Hien Quoc Ngo, Trung Q. Duong and Berk Canberk, "*Secure Full-Duplex Small-Cell Networks in a Spectrum Sharing Environment*", Special Section On Security In Wireless Communications And Networking, July 7, 2016 (DOI 10.1109/ACCESS.2016.2582486)
[40] Mustafa Harun Yılmaz and Hüseyin Arslan, "*A Survey: Spoofing Attacks in Physical Layer Security*", (Invited Paper) 40th Annual IEEE Conference on Local Computer Networks, LCN 2015, Clearwater Beach, Florida, USA
[41] José Marinho, Jorge Granjal and Edmundo Monteiro, "*A survey on security attacks and countermeasures with primary user detection in cognitive radio networks*", EURASIP Journal on Information Security, 11 April 2015 (DOI: 10.1186/s13635-015-0021-0)
[42] Linyuan Zhang, Guoru Ding, Qihui Wu and Fei Song, "*Defending Against Byzantine Attack in Cooperative Spectrum Sensing: Defense Reference and Performance Analysis*", IEEE Access, August 8, 2016 (DOI 10.1109/ACCESS.2016.2593952)
[43] http://www.wipro.com/documents/cognitive-radio.pdf
[44] https://www.xgtechnology.com/wp-content/uploads/2014/07/wp_spectrum_crisis_solution.pdf
[45] https://networks.nokia.com/innovation/5g
[46] Wenmiao Song, "*Configure Cognitive Radio using GNU Radio and USRP*", 2009 3rd IEEE International Symposium on Microwave, Antenna, Propagation and EMC Technologies for Wireless Communications, 27-29 Oct 2009 (DOI: 10.1109/ MAPE.2009.5355934)
[47] Abhijeet Mate, Kuo-Hao Lee and I-Tai Lu, "*Spectrum sensing based on time covariance matrix using GNU radio and USRP for cognitive radio*", 2011 IEEE Long Island Systems, Applications and Technology Conference, 6 May 2011 (DOI: 10.1109/LISAT.2011.5784217)
[48] www.xilinx.com (Xilinx University program)
[49] Universal Software Radio Peripheral (USRP). Information available at http:// www.ettus.com/

IOP Conf. Series: Materials Science and Engineering **247** (2017) 012005 doi:10.1088/1757-899X/247/1/012005

# A Novel Approach to Develop the Lower Order Model of Multi-Input Multi-Output System

Rajalakshmy.P[1]
Assistant Professor
Department of EIE
Karunya University
Coimbatore, TamilNadu, India

Dharmalingam. S[2]
Visiting Professor
Department of ICE
National Institute of Technology
Tiruchirappalli, TamilNadu, India

Jayakumar. J[3]
Professor
Department of EEE
Karunya University
Coimbatore, TamilNadu, India

*Abstract*— **A mathematical model is a virtual entity that uses mathematical language to describe the behavior of a system. Mathematical models are used particularly in the natural sciences and engineering disciplines like physics, biology, and electrical engineering as well as in the social sciences like economics, sociology and political science. Physicists, Engineers, Computer scientists, and Economists use mathematical models most extensively. With the advent of high performance processors and advanced mathematical computations, it is possible to develop high performing simulators for complicated Multi Input Multi Ouptut (MIMO) systems like Quadruple tank systems, Aircrafts, Boilers etc. This paper presents the development of the mathematical model of a 500 MW utility boiler which is a highly complex system. A synergistic combination of operational experience, system identification and lower order modeling philosophy has been effectively used to develop a simplified but accurate model of a circulation system of a utility boiler which is a MIMO system. The results obtained are found to be in good agreement with the physics of the process and with the results obtained through design procedure. The model obtained can be directly used for control system studies and to realize hardware simulators for boiler testing and operator training.**

***Keywords: Mathematical Model, Multi-Input Multi Output (MIMO) system, Lower Order Model, Utility Boiler, Circulation System..***

## I. INTRODUCTION

Modeling and Simulation is a discipline for developing a level of understanding of the interaction of the parts of a system, and of the system as a whole. The level of understanding which may be developed via this discipline is seldom achievable via any other discipline.

A system is understood to be an entity which maintains its existence through the interaction of its parts. A model is a simplified representation of the actual system intended to promote understanding. Whether a model is a good model or not depends on the extent to which it promotes understanding. Since all models are simplifications of reality there is always a trade-off as to what level of detail is included in the model. If too little detail is included in the model one runs the risk of missing relevant interactions and the resultant model does not promote understanding. If too much detail is included in the model the model may become overly complicated and actually preclude the development of understanding. One simply cannot develop all models in the context of the entire universe. Mathematical models can take many forms, including but not limited to dynamical systems, statistical models, differential equations, or game theoretic models. These and other types of models can overlap, with a given model involving a variety of abstract structures.

Mankind has always desired to develop intelligent systems that mimic the biological systems and also obey his commands. The invention of aircraft systems, robots and industrial automations systems and many others has been inspired by this desire. In order to make cost-effective systems for such applications, advancements in computer technology plays an important role. Simulation softwares like Matlab provide a platform to simulate and test the performance of simulator systems before they are implemented on a hardware simulator. This paper deals with the methodology adopted to develop the model of the circulation system of a 500MW utility boiler as a MIMO system. Precise and sustained control of the steam generation system is vital for increasing the boiler efficiency. In order to develop and analyze the control strategies, it is essential to develop an accurate and valid boiler model. A few methods to develop such models are available in literature, but these methods are often plant specific. The objective of this paper is to present a methodology to develop the reduced order model of the circulation system of a boiler.

The behavior of the model described in this paper accurately represents the significant dynamics of the 500 MW utility boiler at Bharath Heavy Electricals Limited, Vishakapatnam, India. Upon review of the currently available models, this paper presents a simple approach to model the circulation system of the 500MW utility boiler based on system identification and lower order modeling philosophies.

## II. MATHEMATICAL MODELING

Mathematical Modelling of any system is important to describe its dynamic behavior. It is a challenging task to accurately model complex real time systems like utility boilers which involve a lot of subsystems. A mathematical model basically establishes the relationship between the input and output of a system. The relationship may be

- Linear or Non-linear Algebraic Equations
- Linear or Non-linear Differential/Partial Differential/Time Varying Equations
- Transfer functions
- State Space Models

Generally, there are two modeling approaches that are adopted in order to determine the dynamic model of a system:

- First Principle or Physical model Approach
- Black Box model Approach

In the physical model approach, the system/process is assumed to be well understood. The modeling philosophy is based on the laws of conservation of mass, energy and momentum as is applicable to the physical process under consideration. This approach usually results in non-linear model equations consisting of differential and algebraic terms. The coefficients that are derived from the model equations have a wide range of validity and are related to the physical process to a great extent.

In the Black Box model approach, the process may not be completely known or may be too complicated to be modeled by physical laws. Hence, the system under study is modeled using its input output relationship. Experiments are performed and field measurements are obtained for input-output responses, from which the process dynamics are determined. This approach is useful for the overall study of various systems.

Many attempts have been made by researchers to derive the lower order model of a thermal power plant. In the case of field measurements, it is found that performing experiments are expensive, time consuming and are prone to error due to measurement noise and the delays introduced due to the sensing devices. Physical model, on the other hand is more reliable and justified since it is derived from fundamental physical laws. Therefore, when such a detailed physical model is available, it is relatively easy to derive the lower order model of the various subsystems or the overall plant. The model thus obtained can be tested for a step or any other disturbance and the response analyzed without having to bother about the complexity involved in the system.

There are many detailed models available in literature for coal-fired thermal power plants. Some of the papers to be mentioned are the ones by Mc Donald and Kwatny [1], Kwan and Anderson [2] in which the authors have derived the physical model of a 100 MW power plant. Reduced order or Lower order modeling of the subsystems in a power plant is necessary because it leads to reduced memory requirements and easy implementation. It can be used to study the effect of controller settings on a system without having to study the detailed mathematical model. It also helps one to realize a hardware simulator that can be used for stability analysis of the power plant and various diagnostic studies.

## III. LITERATURE BACKGROUND

Many attempts have been made by researchers to derive the lower order model of a thermal power plant. Eklund and Gustavsson [3] have used the physical knowledge of the system to derive the structure of the model and identified the transfer function parameters by performing experiments on the boiler. Lower order modeling using mathematical approaches like Newton Raphson method, Jacobian method and many more have been attempted by researchers, but these are found to be computationally complex and time consuming [4 to 20]. Ponnusamy et.al [21] have derived a lower order dynamic model of a complete thermal power plant. Sivakumar and Rajalakshmy have derived a model of a utility boiler furnace using heuristic approach. [22]. In order to derive the lower order model of a system, one needs the input – output measurements obtained from the field or the values generated from the detailed physical model of the plant.

With a detailed physical model is at one's disposal, it is relatively easy to derive the lower order model of the various subsystems or the overall plant. The input-output data can be precisely obtained for a given region of interest from which the lower order models can be derived to fit these input-output relations using appropriate system identification techniques. The model thus obtained can be tested for a step or any other disturbance and the response analyzed without having to bother about the complexity involved in the system.

In contrast with the methods available in literature, the authors have attempted to derive the simplified model of the complex Multiple Input Multiple Output (MIMO) circulation system using the input output data generated from the validated model of a 500 MW thermal power plant at BHEL, Vishakapatnam, India on the basis of literature survey on modeling, plant data knowledge, system identification methodology and heuristic adjustments. The detailed description of the system identification method is discussed and the results are presented.

## IV. SYSTEM UNDERSTANDING

In this paper, a pulverized coal corner fired furnace is considered. There are nine elevations through which coal can be sent to the furnace. For a full load 500MW, fuel is fired through six elevations with appropriate burner tilt. The number of elevations and burner tilt vary depending upon the part load. The net heat input to the furnace is decided by the calorific value and coal flow rate and the sensible heat brought in by primary and secondary air. The feed water from economizer goes to the drum and gets converted into steam in the circulation system (drum – down comers and water walls). The different heat exchangers like panel super heater, platen super heater and re heater which are located at the top of the furnace in the flue gas path. The steam from the drum gets super-heated in various super heaters and enters the high pressure turbine.

The most important subsystems of the thermal power plant are the Boiler Feed Pumps, high pressure boiler (including boiler furnace, Circulation System, Economizer, Super heaters, Reheater) High Pressure, Intermediate Pressure and Low Pressure Turbine, Condenser and De-aerator. The following section describes the methodology and principle behind lower order modeling of the circulation system of a 500 MW coal fired utility boiler. The overall schematic diagram of a boiler system is shown in Fig. 1. [23]

## V.LOWER ORDER MODELING OF CIRCULATION SYSTEM

Reduced order or lower order modeling of various subsystems in a power plant is necessary and useful in a number of situations, because it leads to reduced memory requirements for computation. A low-order model of a power plant has a number of uses:

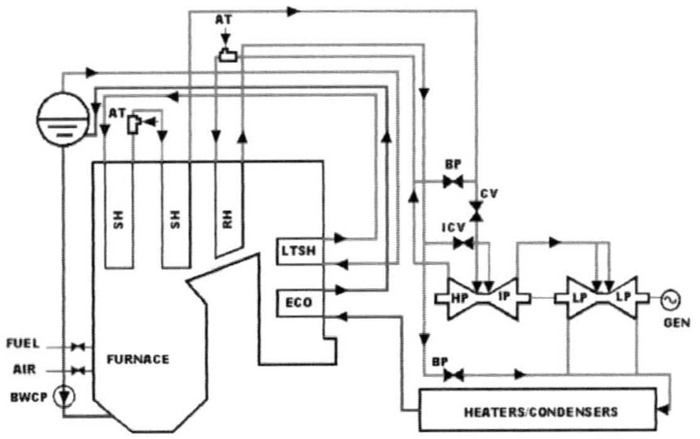


Fig. 1. Overall Schematic diagram of a 500 MW Boiler

- It can be used for various system analysis studies and to study the effect of controller settings on a system without having to study the detailed mathematical model.
- It also helps one to realize a hardware simulator using analog, digital or hybrid hardware.
- It can be used for the study of power plant malfunction or upset conditions within the limits of the model.
- It can be used for long term stability analysis of the power plant and various diagnostic studies.

In summary, it may be inferred that low-order models for the power plant or its subsystems may be derived using input-output responses obtained from field measurements or from the detailed physical model of the power plant, by employing appropriate system identification techniques in each case. Performing experiments are expensive and time consuming and may involve risk to the physical system; hence they are not easy to conduct. Appropriate input signals, such as a step disturbance in only one input, are not easy to introduce. Moreover, in real time situations, all inputs may vary simultaneously even though disturbance was initially introduced only in one of the input variables. Thus in a multi-input, multi-output system, the output recorded would be the result of simultaneous disturbances in several inputs. This makes the identification job more difficult. Also, the field measurements may be corrupted by noise and also be affected by the response lags in transducers and measuring instruments. These factors also have to be taken care of in the identification algorithms. All of these make the generation of input-output data and the subsequent identification process difficult.

This paper adopts the black box model approach and acknowledges the authorities of Bharat Heavy Electrical Limited for permitting access to the input – output responses obtained from the Predicted Performance Data of their Simhadri Power Plant. Predicted Performance of the boiler at different loads will usually be supplied by OEMs. In order to derive the lower order model, the circulation system is considered as a Black Box with multiple inputs and multiple outputs. The system may be considered to behave linearly for small input variations around a steady state operating point. Therefore, the inputs and outputs can be related by suitable transfer functions that are derived from the input-output data of the circulation system. In the proposed methodology, it is observed that the behavior of input and outputs are quite simple and the general form of transfer function can be estimated with ease just by inspection of the input output curves. Once the order is known, then the other parameters like gain, poles and zeros can be easily determined. Software Simulated system identification tool box has been used to estimate the parameters.

The circulation system consists of the drum, down comers and water walls. The feed water that returns from the condenser picks up heat in the High Pressure Heaters and the Economizer before entering the drum at a temperature that is lower than its saturation temperature. Steam that has attained its saturation temperature is drawn out of the drum by adjusting the throttle valve provided at turbine inlet according to the load requirements. The water walls that also form the furnace walls receive heat by direct radiation.

The steps involved in the low-order modeling of the subsystems are obtained and illustrated by considering the circulation system. These low-order models have been implemented in hardware and in a training simulator. The following sections describe the methodology and principle behind lower order modeling of the circulation system of the 500 MW boiler.

The detailed model of the circulation system is quite non-linear. It consists of coupled non-linear differential equations associated with algebraic equations for heat transfer and steam-water properties. These equations can be used to compute the changes in the drum pressure and drum level for any changes in the input variables. Around a given steady state, say the full-load operating point, the incremental changes in the inputs and outputs can be linearly related and the circulation system can be characterized by

$$\begin{bmatrix} G_{11} & G_{21} & G_{31} & G_{41} \\ G_{12} & G_{22} & G_{32} & G_{42} \end{bmatrix} \begin{bmatrix} \Delta T_{weco} \\ \Delta W_{weco} \\ \Delta Q_{gww} \\ \Delta W_{sdro} \end{bmatrix} = \begin{bmatrix} \Delta P_{DR} \\ \Delta V_{DR} \end{bmatrix}$$

Where $G_{xi}(s)$ is the transfer function relating the output at port x to the input at port i.
Thus, the circulation system is perceived as a multi-input multi-output system as shown in Fig. 2.

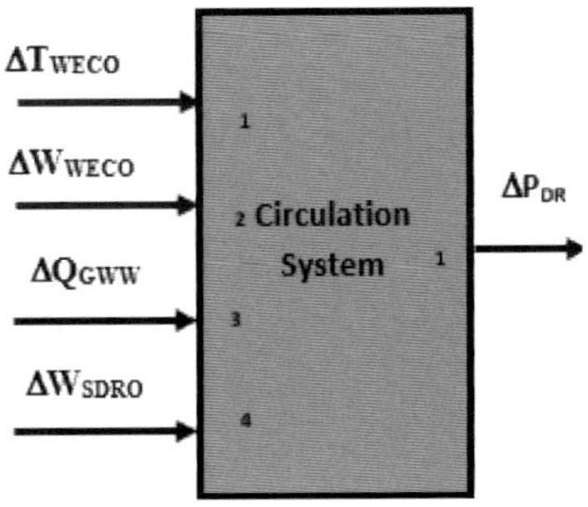


**Fig. 2 Circulation System with the input - output variables**

The schematic representation of the circulation system is shown in Fig.3.[22]
The inputs to the circulation system are

- Feed water Temperature at the Economizer outlet. ($T_{WECO}$)
- Feed water flowrate at the Economizer outlet. ($W_{WECO}$)
- Heat flow rate to water walls. ($Q_{GWW}$)
- Steam flow rate from drum. ($W_{SDRO}$)

The output from the circulation system are

- $P_{DR}$ – Drum Pressure
- $V_{DR}$ – Drum Volume

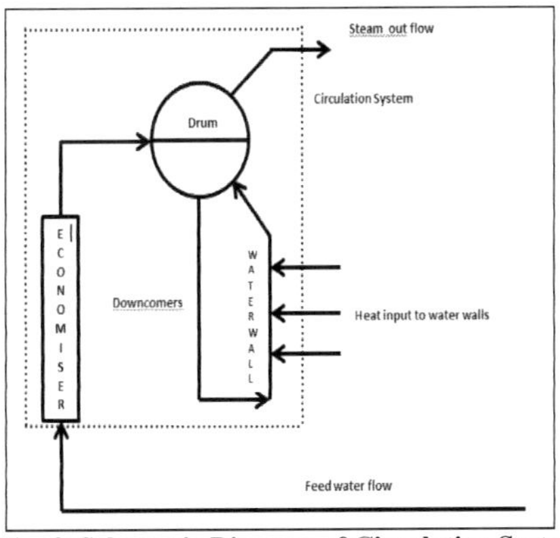


**Fig. 3. Schematic Diagram of Circulation System**

VI. MODEL DESCRIPTION

There exists a detailed non linear mathematical model developed for a 500MW capacity thermal plant in "The Centre of Excellence for Simulators", BHEL, Vishakapatnam, India. [23] This model includes various subsystems such as mills, Primary fans, Secondary fans, Induced draft fans, Boiler feed pumps, furnace and super heating sections as well as reheating sections. The volume and pressure responses of the circulation system model for a step increase in the four input variables are shown in Fig. 4.

This model has been extensively validated from experimental results obtained including field dynamic experiments carried out and used to build plant simulator. The data required for system identification of the circulation system has been generated from this validated model.

The graphs have been plotted against time in seconds. The pressure and volume response for a step increase in the four input variables is observed and a general form of the transfer function is thus estimated.

Now, by using the system identification toolbox, the coefficients are estimated. Fig. 5.a and 5.b. depicts the transfer function representation of the input-output variations.

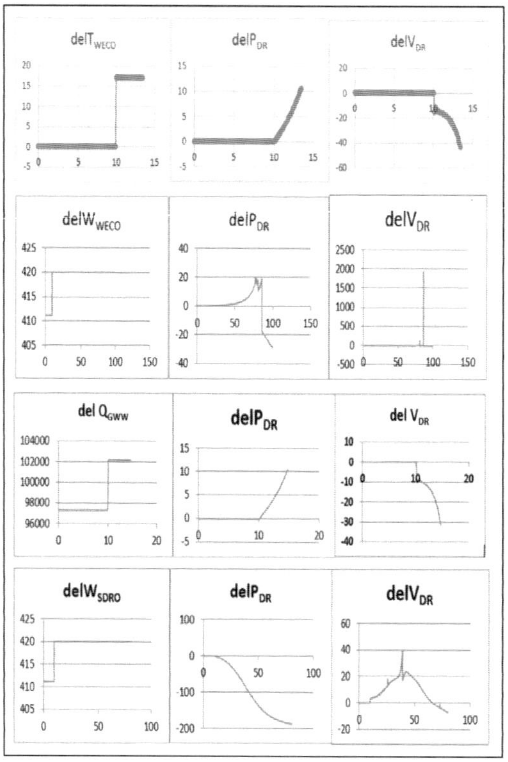

**Fig. 4. Input – Output behavior of a circulation system**

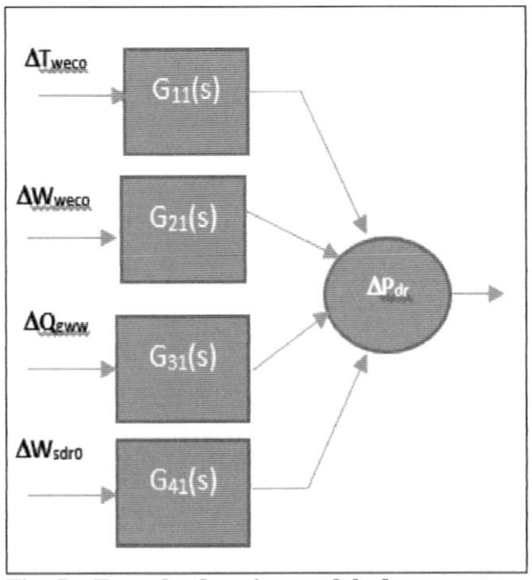


**Fig. 5a. Transfer function model of Input Vs Drum Pressure**

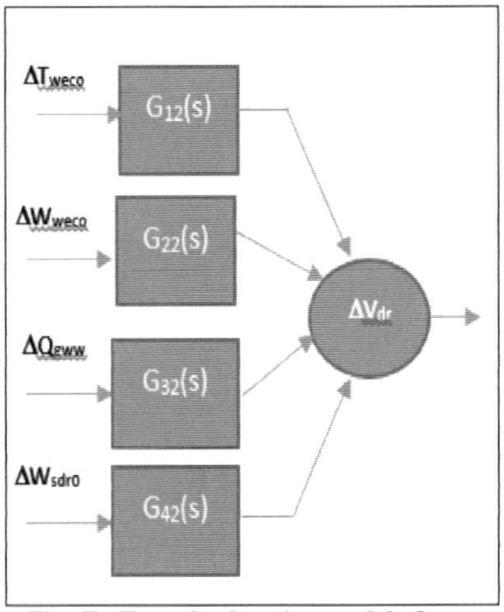


**Fig. 5b. Transfer function model of Input Vs Drum Volume**

The system parameters identified are listed in Table 1 and 2.

**Table 1- System parameters for Input Vs $P_{DR}$**

| S. No. | TF Model | TF Structure | Parameter Identified K |
|---|---|---|---|
| 1 | $G_{11}(s)$ | $\frac{K}{s}$ | 0.017719 |
| 2 | $G_{21}(s)$ | $\frac{K}{s}$ | 0.0013798 |
| 3 | $G_{31}(s)$ | $\frac{K}{s}$ | $2.1007*10^{-6}$ |
| 4 | $G_{41}(s)$ | $\frac{K}{s}$ | $-7.22*10^{-4}$ |

## VII. RESULTS AND DISCUSSION

By observing the response curves of drum Pressure $P_{DR}$ and drum volume $V_{DR}$ for step changes in one of the input variables namely, Wsdro, as shown in Fig. 6a&b, it may be noted that the transfer function relates to the behavior of the system.

The required transfer functions thus obtained can be used for control and simulation studies of the circulation system and further be integrated with the lower order models of the other subsystems to obtain a complete lower order model of a thermal power plant. The transfer functions derived were tested for a step change in the input parameters for drum pressure response and the response obtained is shown in Fig. 7., where $G_{11}$, $G_{21}$, $G_{31}$, $G_{41}$ are the transfer functions derived above.

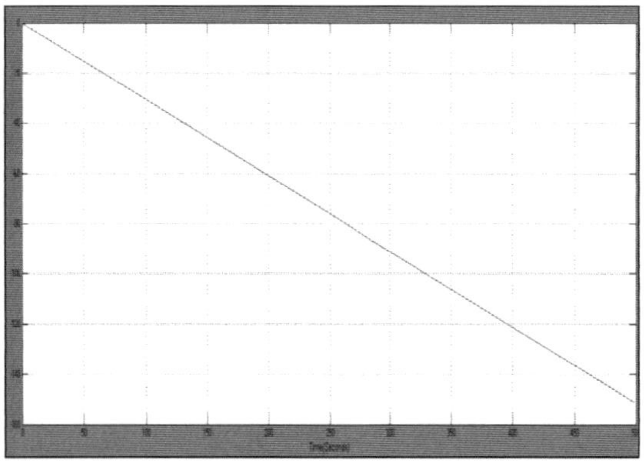

**Fig. 6a. Drum Pressure Response for a step change in Wsdro**

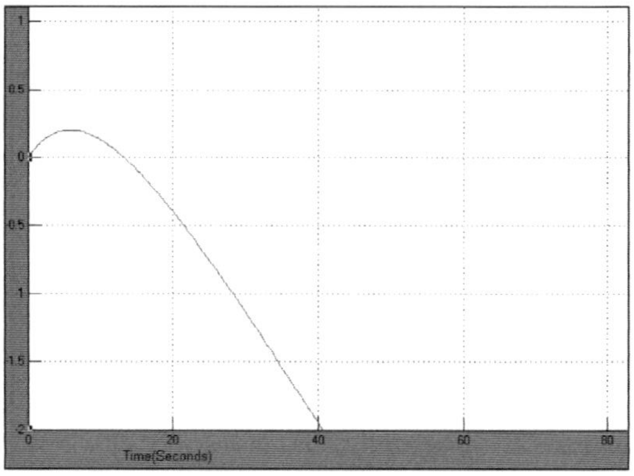


**Fig. 6b. Drum Volume Response for a step change in Wsdro**

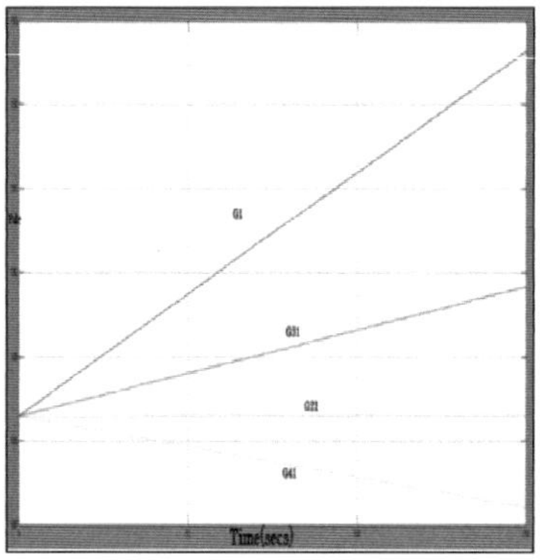


**Fig. 7. Drum Pressure output for a step change in input parameters**

It is observed that the drum pressure response for a step change in the input parameters Tweco, Wweco, Qgww, Wsdro is well within the industrial standards. The simulation has been carried out for a 500MW boiler when the boiler demand is 100%.

The lower order modeling of the complete boiler has been dealt with in two steps. The first step is to develop the mathematical model of the circulation system based on system identification studies. The transfer function obtained is validated by plotting the drum pressure response for varying loads and verifying that they conform to the response of the detailed mathematical model. The response curves of drum pressure $P_{DR}$ for step changes in the boiler input parameters is shown in Fig. 8. It may be noted from the response that the system behavior conforms to the transfer function obtained. For example, the transfer function relating the input step change in the steam flow rate at the outlet of the drum, $\Delta W_{sdro}$ and the output drum pressure, $\Delta P_{DR}$ is defined as

$$\Delta W_{sdro} \frac{\Delta P_{DR}(s)}{\Delta W_{sdro}(s)} = G_{41}(s) = \frac{K_4}{s} .$$

The value of K4 obtained using the system identification technique is -7.22 x 10-4. The negative sign indicates that the step response hence obtained exhibits a negative slope. This behavior of the circulation system conforms to the physical system behavior. This is because the drum pressure decreases for a step increase in the flowrate of steam from the outlet of the drum. In a similar manner, all transfer functions that have been developed in this mode can be tested and validated and the results are quite promising.

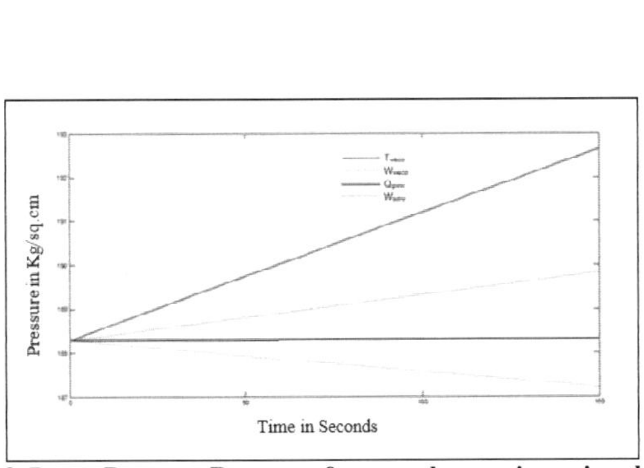


**Fig. 8. Drum Pressure Response for step changes in various loads**

The boiler components are integrated and the entire system is simulated in software for inputs at various loads. The simulation result for 500MW load is shown in Fig. 9. It is the main steam temperature obtained at the output of the platen super-heater

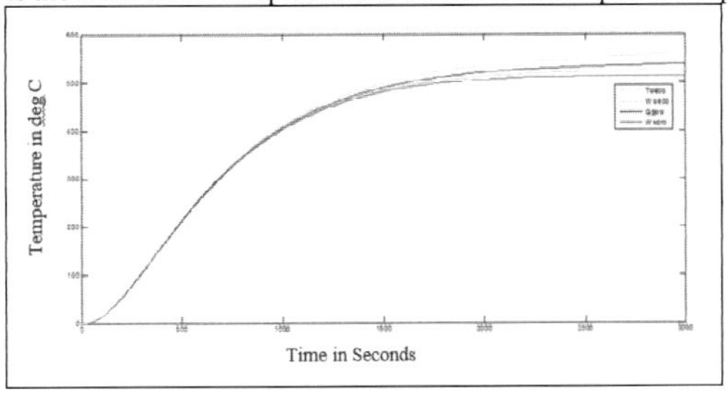


r.

**Fig. 9. Main Steam Temperature Response of Super-heaters for a step change in input variables.**

## VIII.CONCLUSIONS

This paper aims at emphasizing the usefulness of mathematical model in obtaining the lower order model of a circulation system. The required input and output data are accurately generated from the detailed mathematical model of the circulation system and the lower order system parameters are evaluated from these data by employing conventional system identification tool box. The transfer functions obtained can be used for dynamic stability analysis of the system. The methodology can be extended to other MIMO systems like aircrafts, four tank systems as well.

## IX.ACKNOWLEDGEMENTS

The authors wish to express their heartfelt thanks to the authorities of Bharat Heavy Electrical Limited for permitting to access the relevant input-output data from the Simhadri power plant, management of Karunya University, Coimbatore and National Institute of Technology, Tiruchirappalli for providing necessary support.

## REFERENCES

[1] Mc Donald, J.P. ; Kwatny, H.G. "A mathematical model for reheat Boiler-Turbine-Generator Systems", IEEE winter power meeting 70, New York, pp. 1-19, 1970.
[2] Kwan, H. W. ; Anderson, J. H. " A mathematical model of a 200 MW boiler", International Journal of control, Vol. 12, No. 6, pp. 977 – 998, 1970.
[3] Eklund,K. ; Gustavsson,I. "Identification of drum boiler dynamics", Proc. 3rd IFAC symposium on Identification and System Parameter Estimation, Netherlands, Vol.2, pp. 87-108,1973.
[4] Sivakumar, L. ; Bhattacharya, R.K. "Dynamic Analysis of a power boiler using a non linear mathematical model", Proceedings of International Conference of power plant dynamics, Hyderabad, India, pp. 21-29,1979.
[5] Sivakumar, L. ; Reddy, K.L. ; Sundarrajan,N. "Detailed circulation analysis to determine the DNB margin in natural circulation boilers", Proceedings of International conference on Heat and Mass Transfer, Hyderabad, pp. 1-8, 1980.
[6] Profos, P."Dynamic Behaviour of forced flow evaporator systems", Sulzer Technical Review, Vol. 42, No.4, pp.5-12, 1960.

[7] Dolezal, R. ; Varcop, L. " Process Dynamics – Automatic Control of Steam generating plant", Elsevier Publishing Co. 1970.
[8] Laubli ; Fenton. " The flexibility of super critical boiler as a partner in power system design and operation, "Trans. IEEE, Vol. 90, 1971.
[9] Prasada Rao, C. ; Sivakumar, L. " Order and Parameter Identification of continuous linear systems via Walsh functions", Proc. IEEE, pp. 764-766, 1992.
[10] Mathew, A.V. ; Fairman, F.W. "Transfer function matrix identification", IEEE transactions, Circuits and Systems, No.5, pp 584 – 598, 1974.
[11] Dharmalingam, S. ; Sivakumar, L. ; Umapathy, M. "Optimising boiler performance for varying coals using mathematical modeling and embedded controller", 54th ISA POWID Symposium, vol. 487, pp. 298–309, 2011.
[12] Dharmalingam, S. ; Sivakumar, L. ; Anil Kumar, K.K. "Simplified approximations for the accumulation value and time constant for an evaporator system in drum-type boilers", Proceedings of the Institution of Mechanical Engineers, Part A: Journal of Power and Energy, vol. 224, no. 1, pp. 59–67, 2010.
[13] Yadaiah, N. ; Deekshatulu, B.L. ; Sivakumar, L. ; Rao, V.S.H. "Neural network algorithm for parameter identification of dynamical systems involving time delays", Applied Soft Computing Journal, vol. 7, no. 3, pp. 1084–1091, 2007.
[14] Yadaiah, N. ; Sivakumar, L. ; Deekshatulu, B.L. "Parameter identification via neural networks with fast convergence", Mathematics and Computers in Simulation, vol. 51, no. 3-4, pp. 157–167, 2000.
[15] Lixin Tang. ; Huizhi Ren, "Modelling and a segmented dynamic programming-based heuristic approach for the slab stack shuffling problem", Computers & Operations Research, Volume 37, Issue 2, pp 368-375, 2010.
[16] Yong-yao Yang ; Yong-zai Lu. "Dynamic model based optimization control for reheating furnaces", Computers in Industry, Volume 10, Issue 1, pp11-20, 1988.
[17] Gascón-Moreno, J. ; Salcedo-Sanz,S. ; Saavedra-Moreno, B. ; Carro-Calvo, L. ; Portilla-Figueras, A. "An evolutionary-based hyper-heuristic approach for optimal construction of group method of data handling networks", Information Sciences, Volume 247, pp 94-108, 2013.
[18] Anderson, P.M. "Modeling Thermal Power plants for Dynamic Stability Studies", Cyclone copy center, Iowa, USA, 1974.
[19] Chan Ung Park. "Parameter identification of low-order boiler models for dynamic stability analysis", Iowa State University, 1975.
[20] Dharmalingam, S. ; Sivakumar, L.S. ; Anil Kumar, K.K. "Simplified approximations for the accumulation value and time constant for an evaporator system in drum-type boilers", Proc. IMechE Vol. 224 Part A: J. Power and Energy, pp. 59-67, 2011.
[21] Ponnusamy, P. ; Sivakumar, L. ; Sankaran, S. V. "Low-order dynamic model of a complete thermal power plant loop", Proceedings of the Power Plant Dynamics, Control and Testing. Symposium, University of Tennessee, Knoxville, USA, 1, pp. 10.01-10.22, 1983.
[22] Sivakumar.L ; Rajalakshmy.P. " A heuristic approach for modeling of furnace of a utility boiler", IJAER, vol. 10, pp 32723 – 32740, 2015.
[23] Dharmalingam, S. ; Sivakumar, L. ; Anandhi, T. ; Umapathy, M. "Improved method to mitigate the effect of coal fuel switching on Critical Process Parameters of a Steam Generator in a Thermal power plant", Proceedings of the Institution of Mechanical Engineers Part A Journal of Power and Energy 225(8):1026-1040, 2011.

IOP Conf. Series: Materials Science and Engineering **247** (2017) 012006 doi:10.1088/1757-899X/247/1/012006

# AN IMPROVED METHOD TO CONTROL THE CRITICAL PARAMETERS OF A MULTIVARIABLE CONTROL SYSTEM

P. Subha Hency Jims
Assistant Professor,
Department of EIE
Karunya University,
Coimbatore
Tamilnadu, India.
Ph:+91 9194878 46907
hency20002000@karunya.edu

S. Dharmalingam
Visiting Professor,
Department of ICE,
NIT,
Tiruchirappalli
Tamilnadu, India.
Ph: +919443342066
sdharma59@gmail.com

*G. Jims John Wessley
Assistant Professor
Aerospace Engg,
Karunya University
Coimbatore
Tamilnadu, India.
Ph:+9194878406
jims_john@karunya.edu

* Corresponding Author

**Abstract:**

The role of control systems is to cope with the process deficiencies and the undesirable effect of the external disturbances. Most of the multivariable processes are highly iterative and complex in nature. Aircraft systems, Modern Power Plants, Refineries, Robotic systems are few such complex systems that involve numerous critical parameters that need to be monitored and controlled. Control of these important parameters is not only tedious and cumbersome but also is crucial from environmental, safety and quality perspective. In this paper, one such multivariable system, namely, a utility boiler has been considered . A modern power plant is a complex arrangement of pipework and machineries with numerous interacting control loops and support systems. In this paper, the calculation of controller parameters based on classical tuning concepts has been presented. The controller parameters thus obtained and employed has controlled the critical parameters of a boiler during fuel switching disturbances. The proposed method can be applied to control the critical parameters like elevator, aileron, rudder, elevator trim rudder and aileron trim, flap control systems of aircraft systems.

**Keywords:**
Multivariable systems, boiler, PID controller parameters, Load Disturbance

## 1. INTRODUCTION

Power plants using coal as primary fuel is known as fossil fuel power plants (FFPP). Electrical energy plays an important role in determining the quality of life in today's modern world. Coal continues to play a vital role in the generation of electrical energy because of its availability in abundance. The coal properties influence the design and performance of steam generators. The variation in the heating value of coal supplied change the critical process parameters like main steam pressure, main steam temperature etc.. The proportional, Integral and derivative constants of the PID controllers to control the critical parameters of a fossil fuel fired boiler had been calculated for satisfying certain performance requirements during specified disturbances emanating from inputs to the boiler such as variations in calorific value of the coal burnt (fuel switching) in the boiler.

## 2. SYSTEM DESCRIPTION

A thermal power station is a power plant in which the prime mover is steam driven. Coal and air is burnt in the furnace in which the chemical energy is converted to thermal energy. This thermal energy is used to heat water from the boiler drum to turn it to steam and superheat it. The superheated steam is used to drive a steam turbine. In this process, thermal energy is turned into mechanical energy. The steam turbine is coupled to the rotor of an electrical generator, where the mechanical energy is converted to electrical power as shown in Figure 1. After it passes through the turbine, the steam is condensed in a condenser and recycled to where it was heated. The path taken up by the water-steam in this process is called Rankine cycle. The boiler is a drum- and single reheat-type with tilting tangential firing and controlled circulation. The detailed, validated non-linear mathematical model available in the Centre of Excellence for Simulators, BHEL, Hyderabad, India has been utilized for simulation.

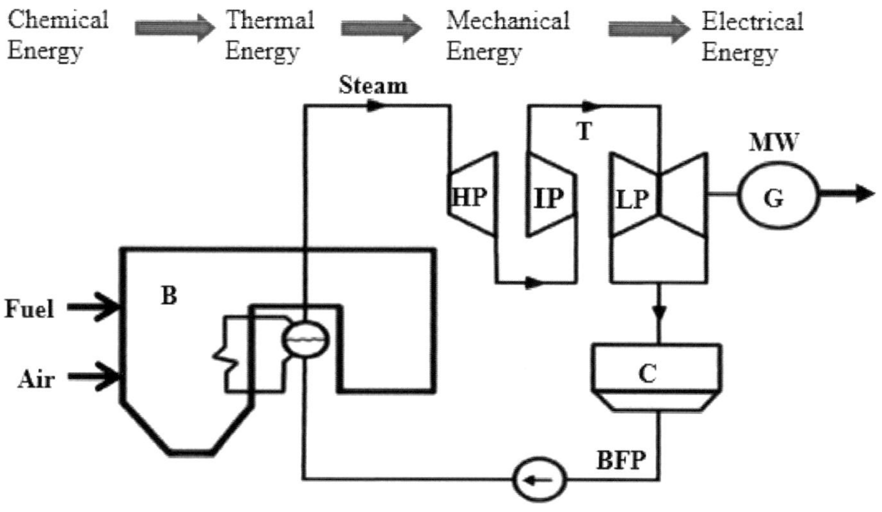


**Fig 1. Energy Conversion in a Fossil Fuel Power plant**

## 3. SIMULATION MODEL OF 500 MW BOILERS AT COE, CORPORATE R&D, BHEL, HYDERABAD

There exist detailed nonlinear mathematical models developed since late 70s by Sivakumar and Bhattacharya (1979), Sivakumar et al.(1980) and Ponnusamy et al. (1983), Sivakumar et al. (1983) and Sivakumar, Sankaran and Mathur (1983) for 120 MW, 210 MW and 500 MW unit sizes in The Centre of Excellence (COE) for Simulators, BHEL, Hyderabad, India.

The overall mathematical model had been developed based on the following modules:

### 3.1 Boiler Pressure parts

- Economizer system module
- Drum, Downcomers and Water Walls (Circulation system) module
- Furnace system module
- Super heater system modules (Low temp super heater, Divisional Panel super heater and Platen super heater)
- Desuper heater system module (in between LTSH and Divisional Panel SH)
- Reheater system module (Reheater Front, Reheater rear)

### 3.2 Boiler Auxiliaries

- Pulverizer (Mill) system module
- Fan system module (Primary air fans, Secondary air fans and Induced draft fans)
- Boiler Feed Pump system module (Two numbers turbine driven boiler feed pumps and one number motor driven boiler feed pump)

### 3.3 Miscellaneous Components

- Throttle Valve
- High Pressure Bypass and Low Pressure Bypass Valves (HPLP Bypass Valves)
- All necessary valves, vanes and dampers.

### 3.4 Control Loops

- Drum Level Controller
- Furnace Pressure controller
- Super heater outlet temperature controller
- Reheater temperature controller
- Master Pressure controller (Fuel flow controller, Total air flow controller with $O_2$ bias)

All the subsystems covered in Section 3 had been modeled by means of first principles approach. Conservation equations for mass, energy and momentum have been considered. Appropriate heat transfer and fluid flow correlations based on collaborators data as well as data from Heat transfer and Fluid Flow (HTFS) documents had been used. All the subsystems mentioned in Section 3.1 and 3.2 were modeled based on equipment performance curves as these are available with BHEL as Original Equipment Manufacturer (OEM). The control loops were synthesized based on ABB P13 hardware and all the logics have been software emulated for integrating control loops with integrated boiler model using various subsystem models obtained as described above.

The model also includes

- For a given coal history (elemental composition), coal flow, secondary air flow, burner tilt and the burner elevations, the Furnace model calculates the following:
  - the flow and the flue gas temperature at the furnace outlet plane.
  - Heat transferred to water walls, platen super heaters, final super heaters and reheaters by direct radiation
- The heat transferred by convection and non-luminous radiation to various heat exchangers such as platen, final and low temperature super heaters, reheaters, economizer and air heaters and thereby the transient variations in pressure, temperature of the inside fluid such as steam, water or air and metal temperature of the heat exchanger.
- Circulation system calculating the drum pressure and water level variations including the effects of swell and shrink.
- Turbine and generator systems, condenser, low pressure and high pressure heaters, de-aerator, condensate pumps and boiler booster pumps.

### 3.5 Model Validation

The integrated model had been tested first as design specific (process variables like pressure, temperature and flow for water, steam, flue gas at different nodes compared with predicted design values and those obtained from models). The error had been less than 0.5 to 1.0%. After the performance guarantee tests on the actual power plant, steady state data for different loads had been collected and the model had been tuned to plant specific. Further, over a period of time, specific open loop tests had been conducted to obtain the transient responses and the model had been further fine-tuned for plant specific. The accuracy limits during transient's zones are well within 5.0 % for wide variety of disturbances.

## 4. BOILER CONTROLS

Automatic control of a process requires a thorough knowledge about the system to be controlled. Also the control of one parameter affects the other parameter under control. Boiler is such a system with highly interactive loops. It is essential that the main steam pressure, main steam temperature, drum water level, furnace pressure be tightly controlled. These parameters are called critical parameters and these controllers are called Class A controllers. PID and PI controllers are the usually employed in controlling the critical process parameters.

a. Master Pressure Control
b. Main Steam Temperature Control

**a. Master Pressure Control**

The objective of this control is to maintain the turbine throttle pressure constant at the desired value by adjusting the firing rate (both fuel flow and air flow). Turbine throttle pressure is measured with primary and redundant transmitters. The measured signal is compared with the set point and any error will have proportional and integral action. A proportional value of total steam flow and derivative of drum pressure signal are taken as feed forward feature for the control. An auto/manual station is provided. The output of A/M station is the air flow demand and fuel flow demand signal.

**b. Main Steam Temperature Control**

One of the critical control loops in a thermal power plant is main steam temperature control. The steam temperature is affected by variations in boiler load. The temperature of saturated steam from boiler drum is increased in super heaters. This temperature if goes beyond a particular limit, reduces the life time of the components of boiler and turbine. The manipulated variable is identified as the spray water flow to control the super heater steam temperature. Through the attemperator, the spray water is injected. It is essential that the control loop is tuned properly to compensate for the variations.

## 5. CLOSED LOOP PERFORMANCE OF THE BOILER DURING LOAD VARIATION

A typical Performance Requirements of Auto Controls for 500 MW boilers for load changes considered for the study is given in Table 1.

**Table 1.Maximum deviation of MS pressure from set point**

| **Load Change(% of TMCR per minute)** | **Allowable deviations in Main Steam Pressure($Kg/cm^2$)** |
|---|---|
| - 3% for five minutes (Load is decreased from 500 MW to 425 MW in 5 minutes) | ± 2 |

The authors have employed the PID controller with controller constants Kp=5.0, Ki=60 and Kd=120 to control the critical parameters.

## 6. SIMULATION RESULTS

Using the mathematical model and tuning the master pressure controller constants to the parameters as above, a small variation in the load was introduced as shown in Fig 2. The boiler demand signal which is the manipulated variable is shown in Fig 3. The response of main steam pressure which is the most critical parameter is shown in Fig.4. The simulation results for various parameters are given in Fig. 5 to 10.

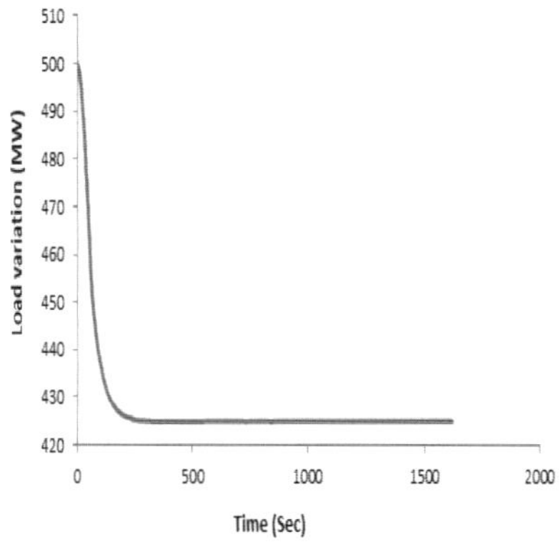


**Fig. 2. Load variation (MW)**

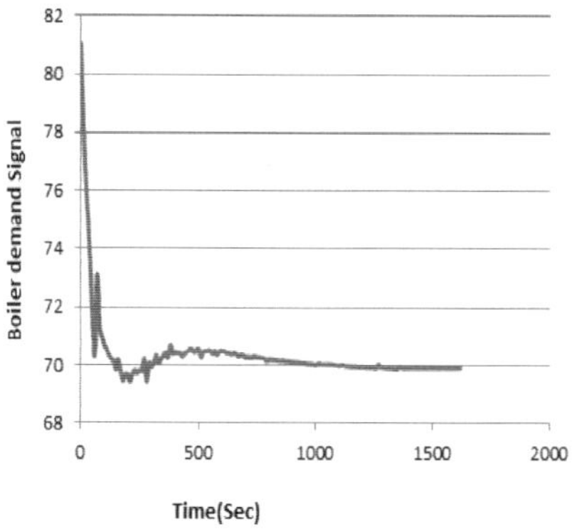


**Fig. 3. Boiler Demand Signal**

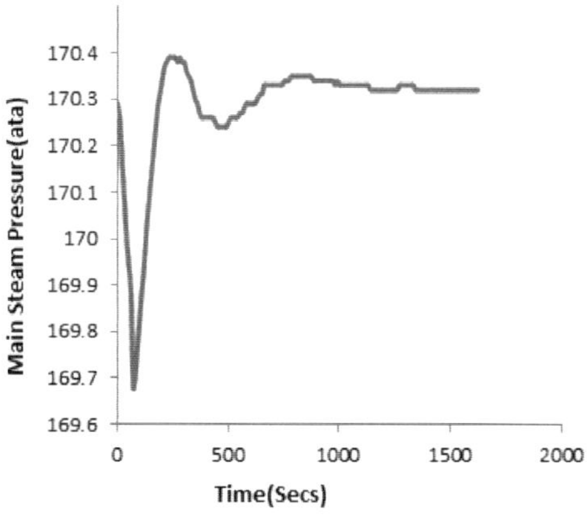


**Fig. 4. Variation in Main Steam pressure**

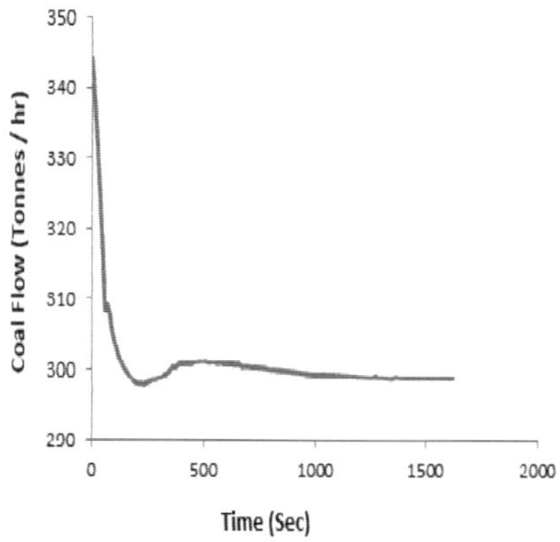


**Fig. 5. Response of Coal Flow Rate**

IOP Conf. Series: Materials Science and Engineering **247** (2017) 012006 doi:10.1088/1757-899X/247/1/012006

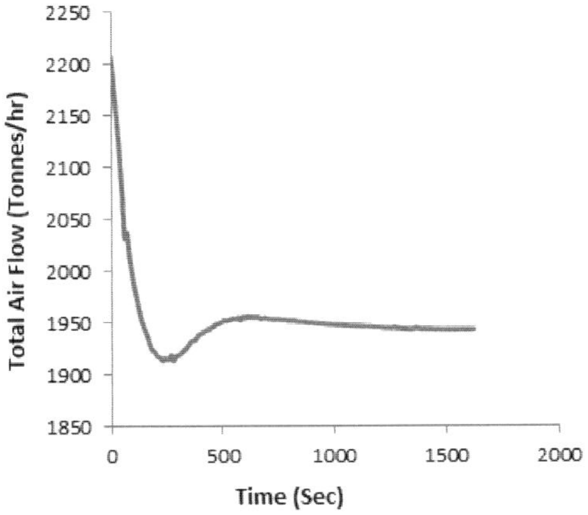


**Fig. 6. Response of Total air flow**

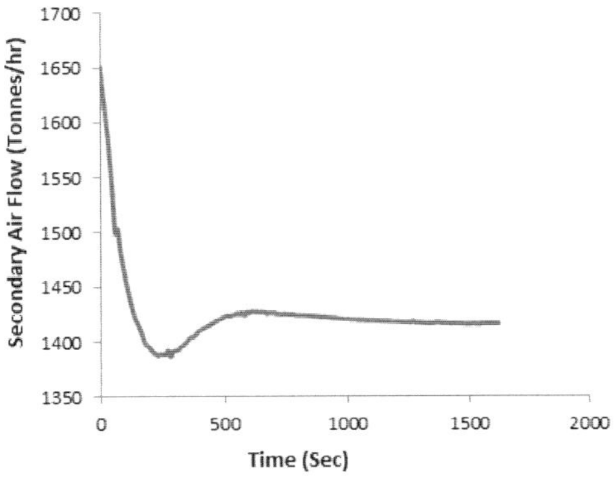


**Fig.7. Response of Secondary Air Flow**

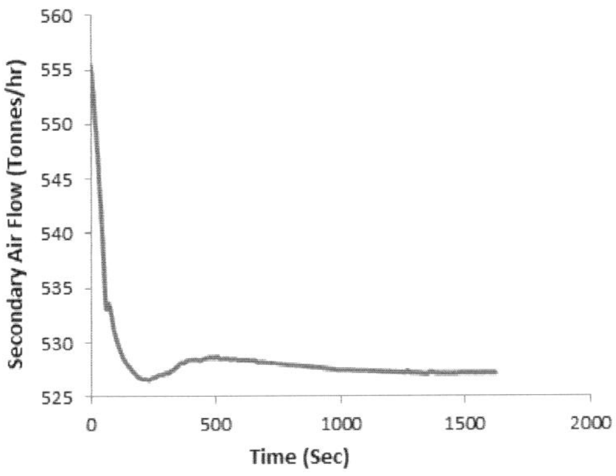


**Fig. 8. Response of Primary Air Flow**

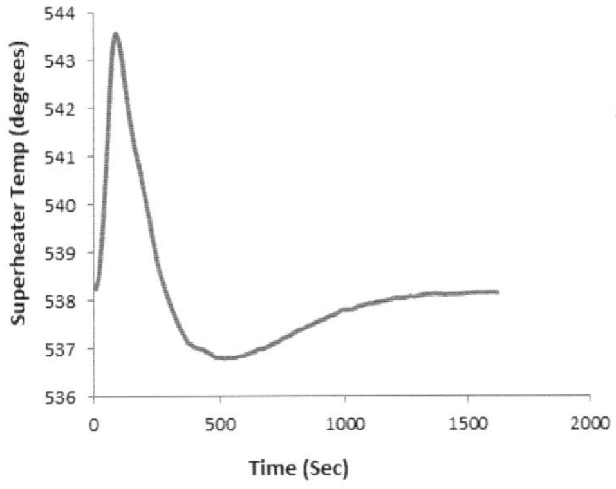


**Fig.9Response of Superheater temperature**

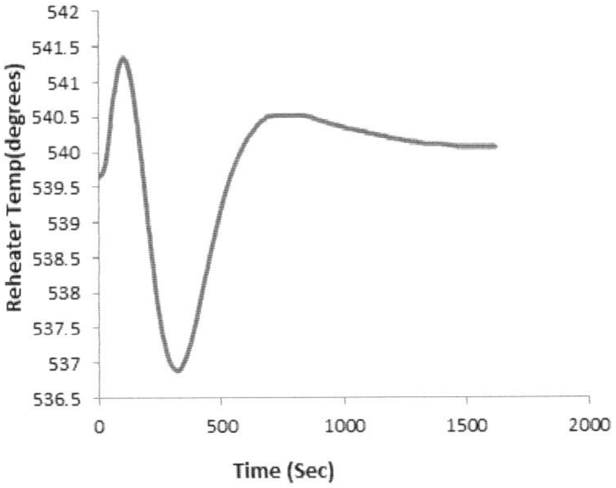


**Fig.10. Response of Reheater Temperature**

It is well noted that all the critical parameters of a 500 MW boiler are controlled by the PID controller. All these parameters are within the limits. The main steam pressure of the 500 MW boiler is well within the allowable deviations. It can be concluded that the PID controller can be applied to any complex systems of aircraft modules.

## 7. CONCLUSION

The dominant tool to understand the transient behavior of the boiler and to develop improved control schemes for coal fired power plants depend on the availability of detailed first principle mathematical model. In this research, the PID controller is employed to satisfy the performance requirements of boiler has been demonstrated. The transient performance of main stream pressure for a change in the load has been obtained. The results prove that the controller parameters employed by PID controller is quite a promising one and can be applied to any complex systems like aircraft systems.

## REFERENCES

[1] Sivakumar, L. and R.K. Bhattacharya (1979) "Dynamic Analysis of a power boiler using a non linear mathematical model." Proceedings of International Conference of power plant dynamics, Hyderabad, India, 21-29.

[2]Sivakumar, L., K.L. Reddy and N. Sundarrajan (1980) "Detailed circulation analysis to determine the DNB margin in natural circulation boilers." Proceedings of International conference on Heat and Mass Transfer, Hyderabad, India, 1-8.

[3]Sivakumar, L., S.V.Sankaran, A. Mathur and A. Dhar (1983) "Mathematical model of (12) subsystems of a 210 MW thermal power plant." Internal reports, Corporate R&D, BHEL, Hyderabad, India.

[4]Sivakumar, L., S.V.Sankaran, and A. Mathur (1983) "TP Plant - A non linearmathematical computer model of a 210 MW thermal power plant." Internal reports, Corporate R&D, BHEL, Hyderabad, India.

[5] Ponnusamy, P., L. Sivakumar and S.V. Sankaran (1983) "Low-order dynamic model of a complete thermal power plant loop". Proceedings of the Power Plant Dynamics, Control and Testing Symposium, University of Tennessee, Knoxville, USA, (1), 10.01-10.22.

[6] P.Rajalakshmy, P.SubhaHency Jose, S, Dharmalingam, S. Jayakumar (2017), :Development of detailed mathematical model based on Chemical and Thermodynamic equations, International journal of Chem Tech Research Vol.10, No.6,2017, pp 202-208.

[7] Dharmalingam S., Sivakumar L. and Umapathy M., :Optimizing boiler performance for varying coals using mathematical modeling and embedded controller. In: 54$^{th}$ Annual ISA POWID Symposium, 6-8 June 2011, Charlotte, NC, USA.

[8] Dharmalingam, S., Sivakumar, L. Anandhi T. and M. Umapathy,: Improved method to

mitigate the effect of coal fuel switching on Critical Process Parameters of a Steam Generator in a Thermal Power plant. In: Proc. IMechE Part A: J. Power and Energy, Vol 225, No.8, 2011 1026-1040.

[9] Dharmalingam, S., Sivakumar, L. and M. Anil Kumar K.K: Simplified approximations for the accumulation value and time constant for an evaporator system in drum-type boilers. In: Proc. IMechE Part A: J. Power and Energy, Vol 224 Part A : J. Power and Energy, pp.59-67.

[10] Dharmalingam. S,:Determination of accumulation value for the evaporator system of fossil fuel Boilers and a novel control system design for fuel switching, Ph.D Thesis, Department of Instrumentation and Control Engineering, NIT, Trichy, July 2011.

[11] L.Sivakumar, Subha hency Jims, S. Dharmalingam, An investigation on the efficacy of classical tuning algorithm to satisfy advanced requirements: control of main steam pressure during fuel switching and load disturbances in coal fired boilers, IFAC-papersOnLine 49-1(2016) 443-449

[12] Astrom, K.J., Hagglund, T,:The Future of PID control. In: Control Engineering Practice 9, 2001 1163-1175.

[13] Astrom, K.J., Hagglund, T,:Revisiting the Ziegler – Nichols step response method for PID control. In: Journal of Process control, 14, 2004, 635-650

[14] Astrom, K.J., Hagglund, T,:PID Controllers: Theory, Design, and Tuning, the Instrumentation, Systems and Automation Society, 2 Sub edition, January 1, 1995, ISBN-10: 1556175167

[15] P. SubhaHency Jims, J. Jayakumar, S. Dharmalingam (2016), "Application of classical step response method to evaluate the PID controller constants for controlling the critical process parameters of a 500 MW steam generator in a thermal power plant during fuel switching", Journal of Electrical Engineering ,Vol.16, 4$^{th}$ Edition, 299-309.

[16] P. Subha Hency Jims, J. Jayakumar, S. Dharmalingam (2017), 'Application of classical step response method to determine the PID controller parameters for a 500 MW Steam Generator in a thermal power plant', Pak. J. Biotechnology, Vol. 14 (Special Issue II) Pp. 242-245.

IOP Conf. Series: Materials Science and Engineering **247** (2017) 012007 doi:10.1088/1757-899X/247/1/012007

# Structural dynamic analysis of turbine blade

**Daniel Antony[1], Gopalsamy M[1], Chaparala B. V. Viswanadh[2], Krishnaraj R[2],**

[1]Assistant professor, [2]UG scholar
Department of Aerospace Engineering
Karunya University
Coimbatore, India

danielantony@karunya.edu, gopalsamy@karunya.edu

**Abstract.** In any gas turbine design cycle, blade design is a crucial element which needs maximum attention to meet the aerodynamic performance, structural safety margins, manufacturing feasibility, material availability etc. In present day gas turbine engines, most of the failures occur during engine development test and in-service, in rotor and stator blades due to fatigue and resonance failures. To address this issue, an extensive structural dynamic analysis is carried out to predict the natural frequencies and mode shapes using FE methods. Using the dynamics characteristics, the Campbell diagram is constructed to study the possibility of resonance at various operating speeds. In this work, the feasibility of using composite material in place of titanium alloy from the structural dynamics point of view. This is being attempted in a Low-pressure compressor where the temperatures are relatively low and fixed with the casings. The analysis will be carried out using FE method for different composite material with different lamina orientations chosen through the survey. This study will focus on the sensitivity of blade mode shapes to different laminae orientations, which will be used to alter the natural frequency and tailor the mode shapes. Campbell diagrams of existing titanium alloy are compared with the composite materials with different laminae at all critical operating conditions. The existing manufacturing methods and the proven techniques for blade profiles will also be discussed in this report.

## 1. Introduction

For many decades, researchers have steadily worked on improving the performance of gas turbine engines. Major advances were made in the areas of materials, structures, controls, reliability, thermal cycles, thrust to weight ratio, and overall pressure ratios. Improving performance, reducing operating cost, lowering emission and noise, and enhancing structural reliability of gas turbine engines are prudent to meet the challenges of the new century. Production of lighter engines than conventional engines with higher efficiency and improved reliability compared to present technology is essential. composites are versatile materials have so far been best known for their weight saving contribution to nacelles, thrust reversers, ducts, guide vanes and other items outside the engine proper. Polymer composites are temperature limited and therefore unsuitable for use in engine hot sections but not in case of cold sections. Designers consider most benefit in the large fan blades i.e. Low-Pressure compressor blades, found at the front of turbofan engines that power the airliners of today.

Weight reduction is always a strong driver in aeronautics and a set of fan blades that is lighter than any that existed hitherto will automatically help in producing lighter engines that reduce fuel burn while increasing payload and range. However, the dynamics of a fan multiply the weight-saving significance several times over. Blades on fans in current generation turbofan engines can experience centrifugal loadings of around 100 tons-equivalent to the weight of a diesel locomotive hanging on each blade. Typical military gas turbines will revolve at 10,000 rpm in cruise for LP Compressor Spool and the centrifugal load experienced by the rotor blades increase with centre of gravity location, rotational speed and with blade mass. Radius and speed are determined by aerodynamic design considerations, but anything the designers can do to reduce blade mass will reduce loadings hence it is advantageous to change from conventional titanium alloy blades to composites and sandwich structures. This provides over 35-40% weight reduction and also provides good dynamic characteristics.

## 2. Scope of the project

The scope of the project is to carry out the modal analysis of a trial blade profile which may be used in gas turbines by applying composite materials including pre-stressed effects and to find out the natural frequencies and mode shapes. To plot the Campbell diagram and to predict the occurrence of resonance for different speeds. To achieve the above, the following objectives have to be carried out Selection of the best composite material by carrying out the modal analysis of different type of composite materials in a rectangular plate and best orientation of fibres in the plate. Analysing the blade profile available for dynamic characteristics using the available FEM software like ANSYS. Identify the occurrence of resonance, if any at operational conditions and make require changes in the design phase of the structure of the component so as to avoid the resonance at different operating conditions.

### *2.1. FEM in ANSYS*

The finite element method (FEM) is the most popular simulation method to predict the physical behaviour of systems and structures. Since analytical solutions are in general not available for most daily problems in engineering sciences numerical methods have been evolved to find a solution for the governing equations of the individual problem. Although the finite element method was originally developed to find a solution for problems of structural mechanics it can nowadays be applied to a large number of engineering in which the physical description results in a mathematical formulation with some typical differential equations which can be solved numerically.

### *2.2. Static analysis*

Static analysis is used to calculate the effects of steady loading conducting on a structure, while ignoring inertia and damping effects, such as those caused by time varying loads. Static analysis is to determine displacements, stresses, strains and forces in structures or components caused by loads that do not include significant inertia and damping effects.

### *2.3. Dynamic analysis*

Under the dynamic analysis comes Vibration, free vibration, forced vibration, natural frequency, modes, mode shapes, resonance, pre-stressed modal analysis

### *2.4. Campbell diagram*

In rotating machinery, as the structure undergoes rotation the centrifugal load on the blades varies with RPM along with the associated aerodynamic parameters such as pressure and temperature. Due to the variation in the temperature and centrifugal loads, the structural stiffness changes which ultimately results in the change in natural frequency. This change in natural frequency needs to be predicted at the design stage and the possibility of any resonance need to be studied and avoided, if possible.

Mainly two types of excitation namely: Mechanical and Aerodynamic.

Mechanical Excitation: Due to unbalance force in terms of mass rotating with its centre of gravity i.e. U= m*e*ω2, which varies with speed is always a dominating excitation force in rotating machinery. This force will invariably present in the structure and produces an excitation force. The excitation frequency will be equal to the first harmonic of rotor speed i.e. 1X = w /60 * 1. Similarly, other harmonics arise due to excitation frequencies corresponding to looseness, circumferential ovality and so on. Each harmonics of rotor speed has specific relation to a mechanical phenomenon and explained in detail in the following reference [1].

Aerodynamic Excitation: The fluid passing over the structural members like blades, vanes and supporting frames do expand completely and results in a smooth flow. However, the vortices leave the structural member at the trailing edge of the aerofoil and exerts a pulsating force which impinges on the next stage. This again produces an excitation frequency corresponding to the no.of disturbances seen in one revolution. For example, if the no of stator blades in a stage corresponds to 47, then the next following rotor blade will face 47 pulsations in one revolution. The excitation frequency corresponding to that will be 47*RPM/60 and this is termed as 47X in a Campbell diagram. This gives the possibility of resonance to the structural member where the excitation frequency matches with the natural frequency. This information can be inferred from the Campbell diagram (In Fig 3.).

where the variation of natural frequency (Y axis) is plotted against speed (X-axis) along with the engine order lines (i.e. 1X. 2X....47X...so on) applicable to the stage. From the Campbell diagram, it is possible to identify the possibility of resonance at higher RPM's or at regimes where the engine will dwell for longer time and it can be avoided at the design stage.

### *2.5. Pre-stressed modal analysis*

First, the modal analysis for 0 rpm is carried out in the structure using Block Loncoz method and the natural frequencies are noted down. Then, the inertial load is given and a static analysis is carried out with considering the pre-stressed effects due to the inertial load. Then finishing that static analysis, the modal analysis is carried out with switching on the include pre-stress effects option and the natural frequencies are obtained.

## 3. ANALYSIS OF TRIAL BLADE PROFILE

### *3.1. INTRODUCTION*

Modal Analysis is carried out for the given Trial blade profile by applying Composites and comparing its Frequency values with the conventional Titanium blade. The blade profile is shown below:

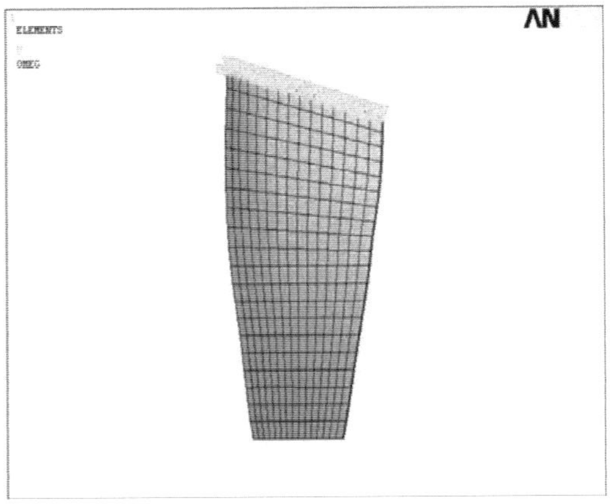

**FIG 1** Front view of trail blade profile

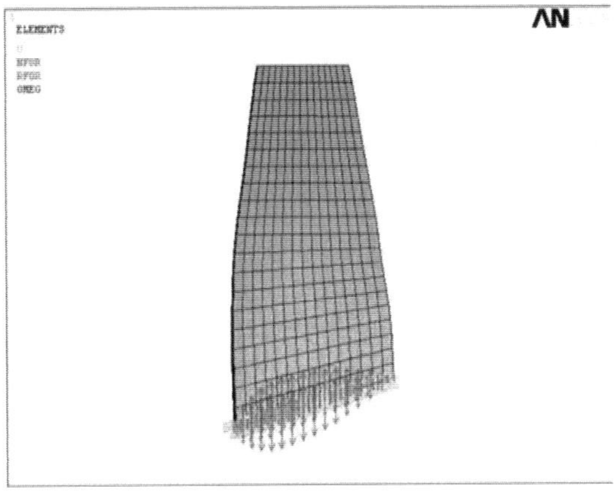

**FIG 2** Front view of trail blade profile with pre-stress effects

### *3.1.1 ANALYSIS OF BLADE PROFILE WITH TITANIUM*

The Blade profile is analysed with Titanium material with following material properties

**Table 1.** Titanium properties

| MATERIAL | YOUNG'S MODULUS, E (Gpa) | DENSITY ρ (Kg/m3) | POISSON RATIO 'Y |
|---|---|---|---|
| **Titanium** | 110 | 4500 | 0.3 |

The table below gives the Frequency values of first four modes of the given blade profile for various Rpm with the help of Pre-stress effects.

IOP Conf. Series: Materials Science and Engineering **247** (2017) 012007 doi:10.1088/1757-899X/247/1/012007

**Table 2.** Frequency values for titanium at different RPM

| MATERIAL | 0 Rpm<br>Frequency (Hz) | 7500 Rpm<br>Frequency (Hz) | 9000 Rpm<br>Frequency (Hz) |
|---|---|---|---|
| Titanium | 455.42<br>1182.6<br>1469.5<br>2233.7 | 601.32<br>1326.6<br>1530.2<br>2265.4 | 680.37<br>1401.9<br>1581.3<br>2271.7 |

*3.1.2 MODES SHAPES OBTAINED WITHOUT PRE-STRESS EFFECTS*

The first three mode shapes of the Titanium blade profile without pre-stress effects are given below,

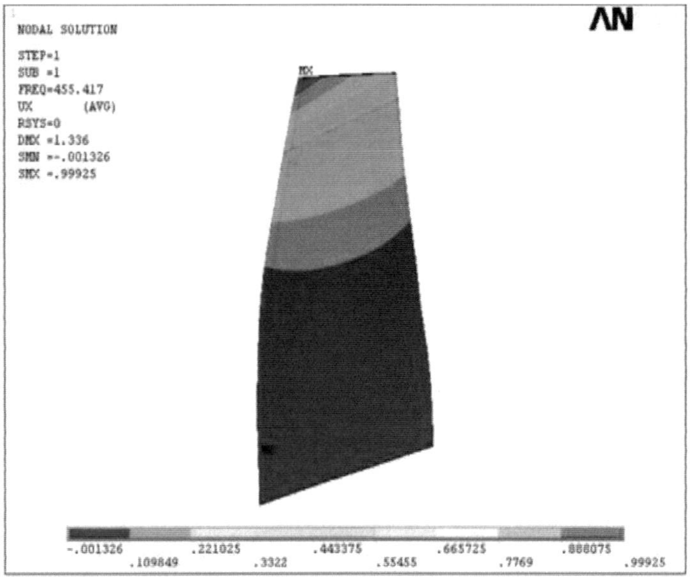


**FIG 3** First mode shape of titanium without pre-stress effects

In the first mode, the maximum displacement will be at the tip and at the root there is minimum deformation

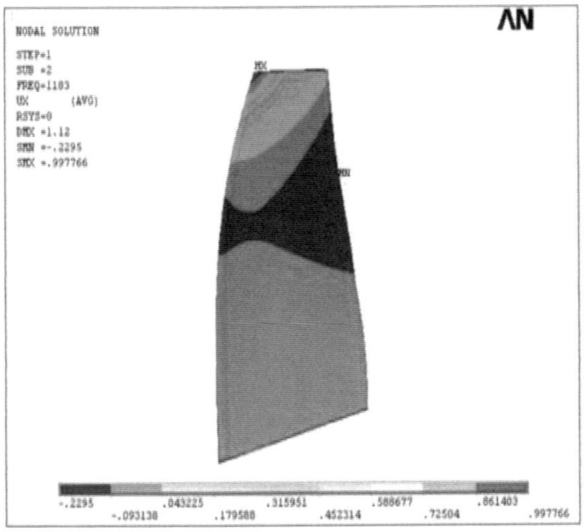


**FIG 4** Second mode shape of titanium without pre-stress effects.

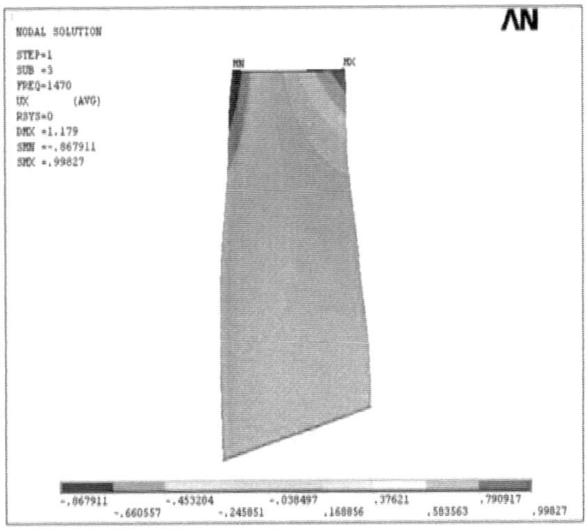


**FIG 5** Third mode shape of titanium without pre-stress effects.

In the second mode, the deformation will be at both the root tip and there will be minimum deformation (-.2295 m) at the middle of the blade profile. In the third mode the deformation will be all over the blade

*3.1.3 MODE SHAPES OBTAINED WITH 7500 RPM*

The First three mode shapes of Titanium blade profile with 7500 Rpm are given below,

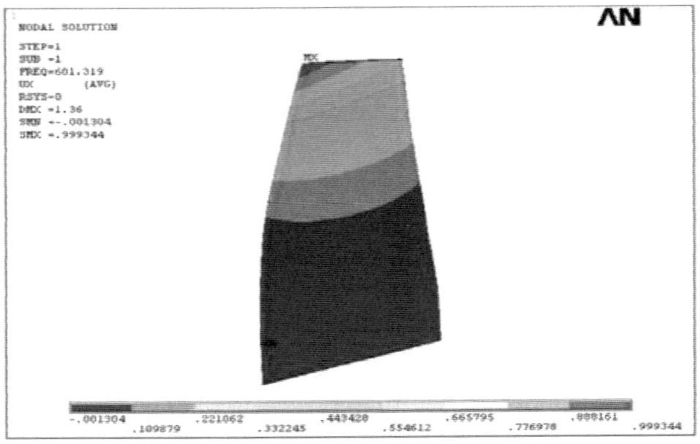


**FIG 6** First mode shape of titanium with 7500 Rpm

In the first mode shape with 7500 rpm, the deformation is similar to mode shape with pre-stress effect

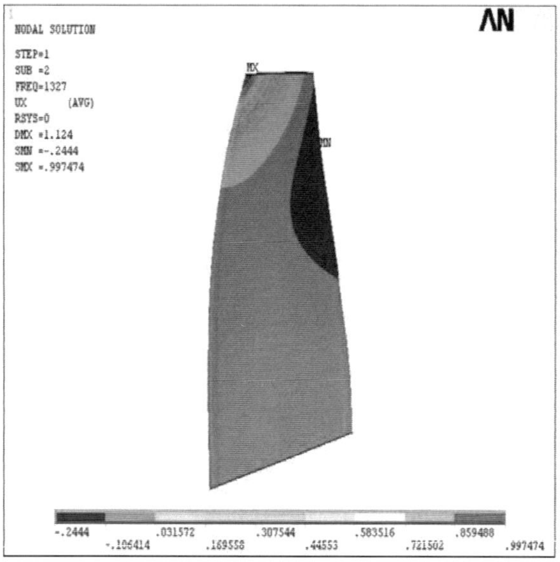


**FIG 7** Second mode shape of titanium with 7500 Rpm

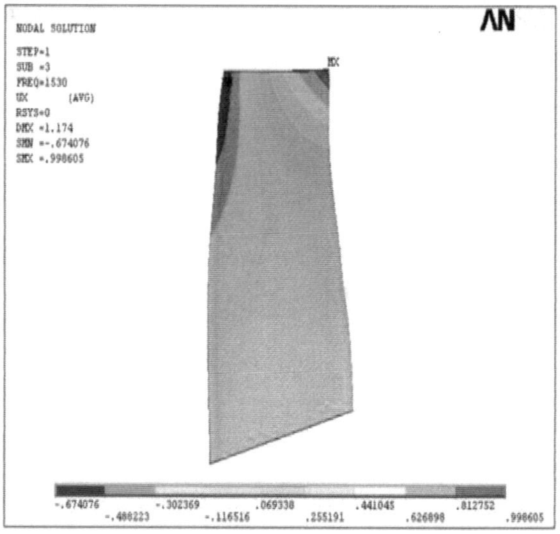


**FIG 8** Third mode shape of titanium with 7500 Rpm

In the second mode with 7500 rpm the deformation will be minimum at the trailing edge and maximum at leading edge of the tip. In the third mode with 7500 rpm, deformation will be all over the blade profile but when compared to mode shape of pre-stress effect the mode shape is varied slightly

*3.1.4 MODE SHAPES OBTAINED WITH 9000 RPM*

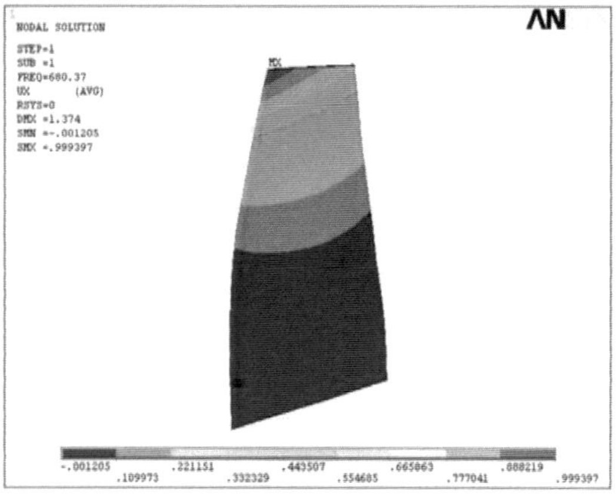


**FIG 9** First mode shape of titanium with 9000 Rpm

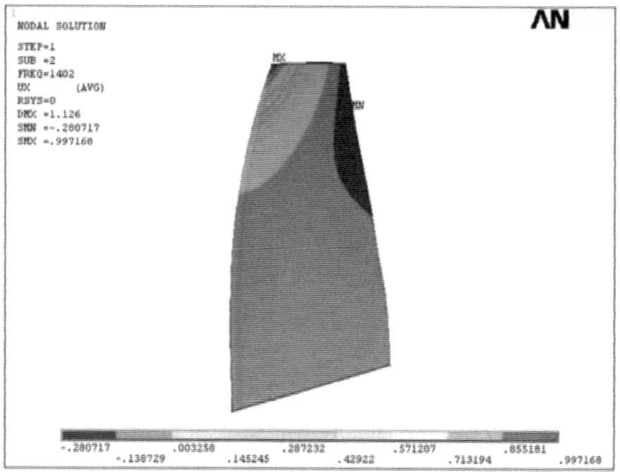


**FIG 10** Second mode shape of titanium with 9000 Rpm

IOP Conf. Series: Materials Science and Engineering **247** (2017) 012007 doi:10.1088/1757-899X/247/1/012007

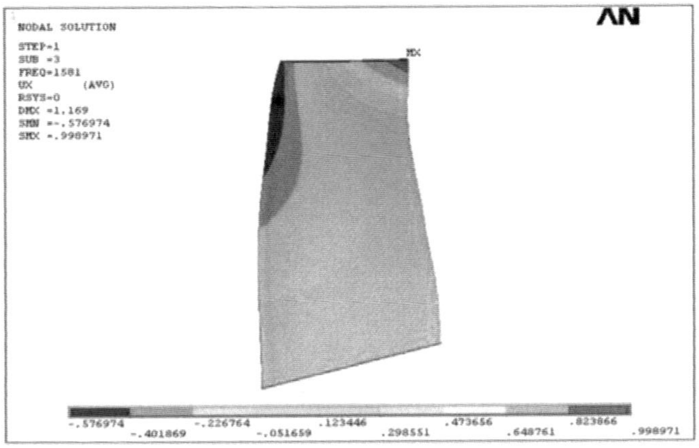


**FIG 11** Third mode shape of titanium with 9000Rpm

All the mode shape found at 9000 rpm are similar to the mode shapes found at 7500 rpm, which are shown above, but the frequency values only varies

### *3.1.5 CAMPBELL DIAGRAM FOR TITANIUM*

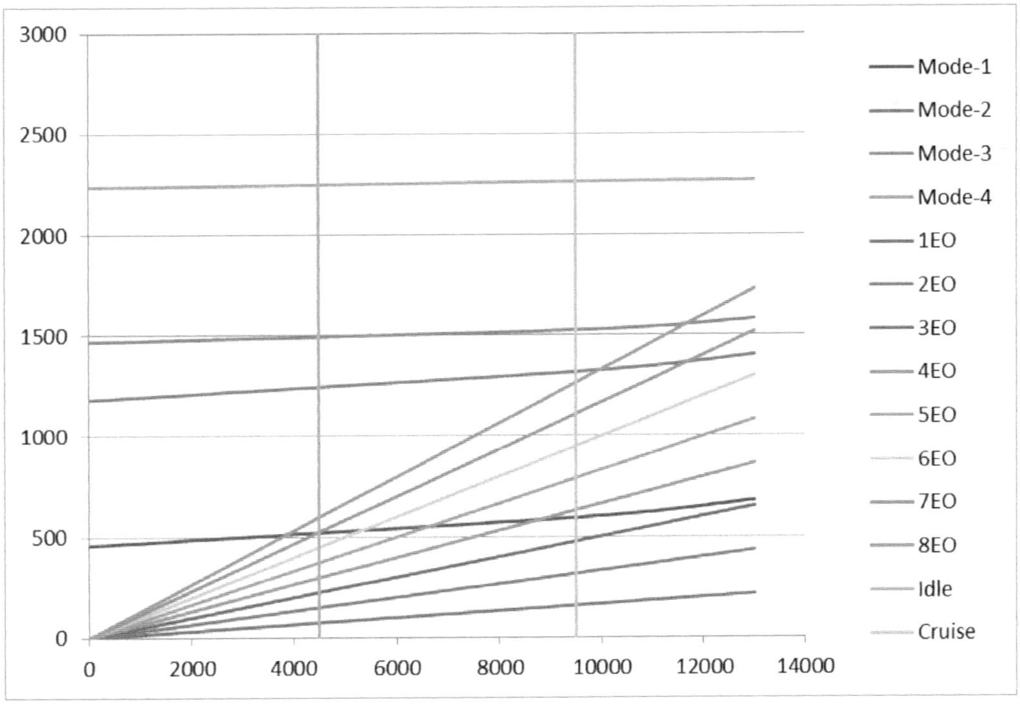


**FIG 12** Campbell diagram for titanium blade

*3.2 ANALYSIS OF BLADE PROFILE WITH THREE LAMINA COMPOSITE MATERIAL*

As a pre-step of applying best Composite material in give blade profile, study was done on four different composite materials which are given below,

- Carbon epoxy
- Graphite epoxy
- Glass epoxy
- Aramid epoxy

Based on Structural Dynamics point of view, Modal analysis was performed for all the above Composite material and from the obtained frequency values Campbell diagram was generated, and as a result best Composite material was chosen. The material properties of lamina were calculated with the help of Micromechanics approach.

The material properties of above composite materials are given below,

**TABLE 3** Material properties of composite materials

| COMPOSITES | E1 (In GPa) | E2 (In GPa) | Nu12 | Nu21 | G12 (in GPa) | DENSITY (Kg/m3) | MASS (Kg) |
|---|---|---|---|---|---|---|---|
| **Carbon epoxy** | 124.83 | 10.46 | 0.3 | 0.025 | 3.33 | 1512 | 0.1007 |
| **Graphite epoxy** | 150.69 | 9.45 | 0.3 | 0.019 | 3.365 | 1590 | 0.1058 |
| **Glass epoxy** | 56.44 | 9.04 | 0.235 | 0.03764 | 3.497 | 2045 | 0.1362 |
| **Aramid epoxy** | 81.79 | 9.243 | 0.339 | 0.03831 | 2.065 | 1330 | 0.0885 |

IOP Conf. Series: Materials Science and Engineering **247** (2017) 012007 doi:10.1088/1757-899X/247/1/012007

The Modal analysis for above four Composite material with different Orientation for Three Lamina is analysed. From the Modal Analysis of above different composite materials, the Frequency values are tabulated as shown,

**TABLE 4** Frequency values of different composite materials for Three Lamina

| Ply | Carbon | Graphite | Aramid | Glass |
|---|---|---|---|---|
| **0/0/0** | 242.38 | 225.46 | 240.05 | 194.85 |
| | 613.06 | 573.8 | 593.96 | 497.02 |
| | 804.8 | 766.44 | 745.94 | 666.07 |
| | 1236.9 | 1183.2 | 1122.6 | 1020.4 |
| **30/45/60** | 299.64 | 293.95 | 271.05 | 229.9 |
| | 760.08 | 739.82 | 692.23 | 590.89 |
| | 981.92 | 961.09 | 880.91 | 770.34 |
| | 1396.9 | 1347.5 | 1279.8 | 1110.7 |
| **45/60/90** | 385.04 | 384.19 | 339.16 | 287.06 |
| | 944.27 | 936.12 | 843.79 | 707.62 |
| | 1150.3 | 1144.6 | 1014 | 866.24 |
| | 1614.6 | 1605.2 | 1431.8 | 1210.1 |
| **90/90/90** | 679.75 | 704.96 | 585.56 | 436.07 |
| | 1135.7 | 1135 | 994.15 | 839.61 |
| | 1534.8 | 1541.2 | 1327 | 1087.4 |
| | 1654.7 | 1669.9 | 1450 | 1220.1 |

From the table that carbon epoxy and aramid epoxy has good dynamic characteristics.

So, further study on lamina orientation is analyzed on carbon epoxy and aramid epoxy as shown,

IOP Conf. Series: Materials Science and Engineering **247** (2017) 012007 doi:10.1088/1757-899X/247/1/012007

**TABLE 5** Frequency values of carbon &aramid composite materials

| orientation | carbon epoxy | Aramid epoxy |
|---|---|---|
| 0/45/0 | 251.02 | 245.69 |
| | 647.53 | 621.93 |
| | 904.39 | 828.36 |
| | 1327.8 | 1233.3 |
| 45/0/45 | 268.52 | 243.11 |
| | 732.68 | 669.98 |
| | 1063.1 | 976.07 |
| | 1487.6 | 1371.1 |
| 0/90/0 | 297.97 | 279.91 |
| | 743.37 | 690.01 |
| | 1088.2 | 943.7 |
| | 1406.5 | 1236.3 |
| 90/0/90 | 672.86 | 578.78 |
| | 1160.5 | 1000.4 |
| | 1544.9 | 1323.6 |
| | 1693.1 | 1476.4 |

*3.2.1 COMPOSITE BLADE WITH 3 LAMINA ORIENTATION*

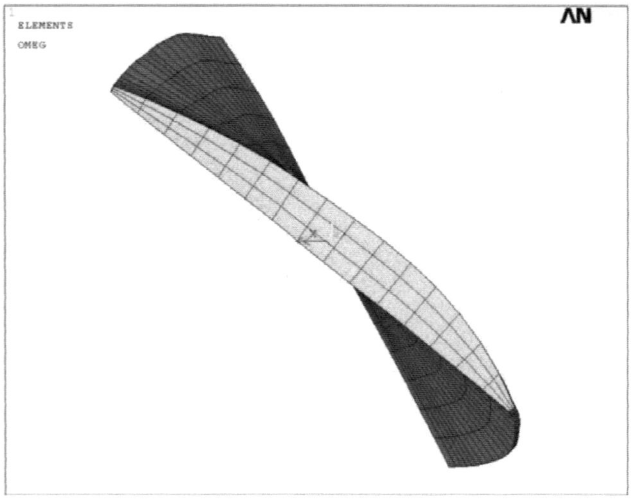


**FIG 13** Composite blade with 3 laminas

IOP Conf. Series: Materials Science and Engineering **247** (2017) 012007 doi:10.1088/1757-899X/247/1/012007

From the **TABLE 5** above it is evident that Carbon Epoxy exhibits its good Dynamic characteristics. Also, Carbon Epoxy has high temperature regime compared to aramid epoxy. So now carbon epoxy is analyzed with pre-stress effects (i.e. with RPM of 7500 and 9000).

**TABLE 6** Frequency values of carbon epoxy for different Rpm

| Orientation | 0 Rpm Frequency(Hz) | 7500 Rpm Frequency(Hz) | 9000 Rpm Frequency(Hz) |
|---|---|---|---|
| **0/45/0** | 251.02 | 456.93 | 548.49 |
| | 647.53 | 871.06 | 940.48 |
| | 904.39 | 1009.7 | 1121 |
| | 1327.8 | 1519.4 | 1607.3 |
| **45/0/45** | 268.52 | 470.76 | 562.85 |
| | 732.68 | 971.28 | 1094.1 |
| | 1063.1 | 1121.6 | 1162.2 |
| | 1487.6 | 1703.2 | 1729.2 |
| **0/90/0** | 297.97 | 485.95 | 573.72 |
| | 743.37 | 969.68 | 1071.8 |
| | 1088.2 | 1150.3 | 1211.4 |
| | 1406.5 | 1529.6 | 1593.7 |
| **90/0/90** | 672.86 | 770.86 | 829.74 |
| | 1160.5 | 1210.4 | 1239.1 |
| | 1544.9 | 1623.2 | 1635.3 |
| | 1693.1 | 1709.2 | 1741.1 |

From the above table, it is evident that Carbon Epoxy shows very good Dynamic characteristics for 90/0/90 ply orientation.

### *3.2.2 MODE SHAPES OF THREE LAMINA CARBON EPOXY WITHOUT PRE-STRESS EFFECT FOR 90/0/90 ORIENTATION*

From the Modal Analysis of three lamina Carbon Epoxy without pre-stress, the first 3 mode shapes are extracted as shown,

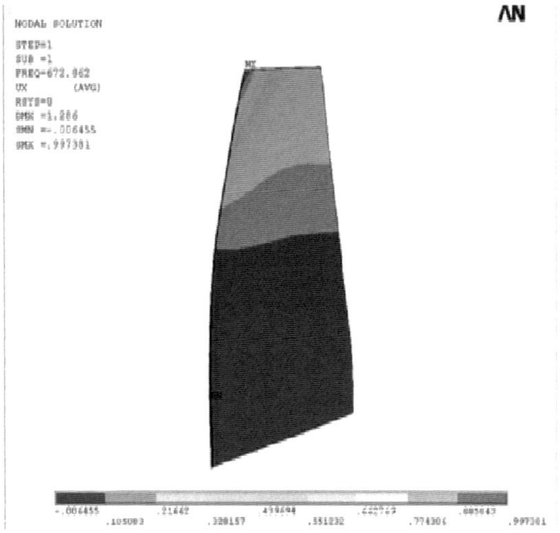


**Fig 14** First mode shape of carbon epoxy for (90/0/90) orientation

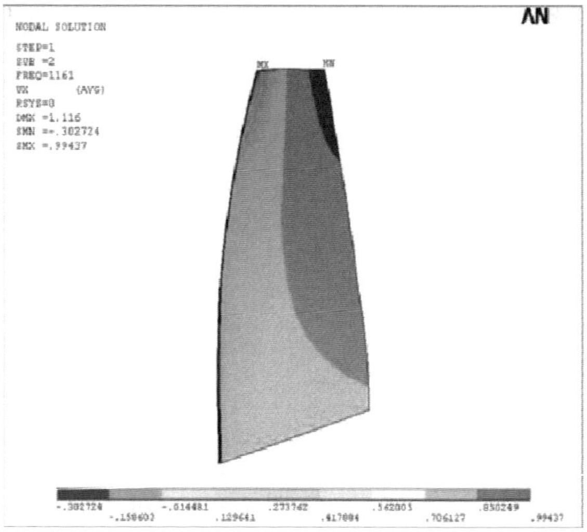


**Fig 15** Second mode shape of carbon epoxy (90/0/90) orientation

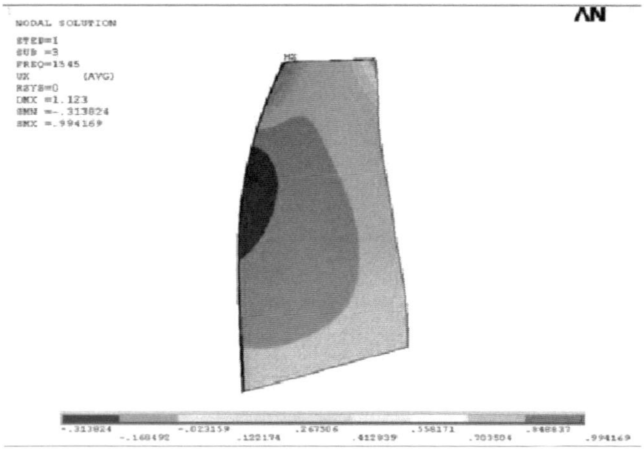


**Fig 16** Third mode shape of carbon epoxy for (90/0/90) orientation

In the first mode shape the maximum displacement will be at the leading edge of the tip of blade while the minimum displacement will be at the root of the blade. In the second mode shape the maximum displacement will be at leading edge of the tip while the minimum displacement will be at the trailing edge of the tip. In the third mode shape, maximum displacement will be at both the leading and trailing edge of tip, while the minimum displacement will be at the middle portion of the leading edge of the blade.

*3.2.3 MODE SHAPES OF THREE LAMINA CARBON EPOXY WITH 7500RP M FOR 90/0/90 ORIENTATION*

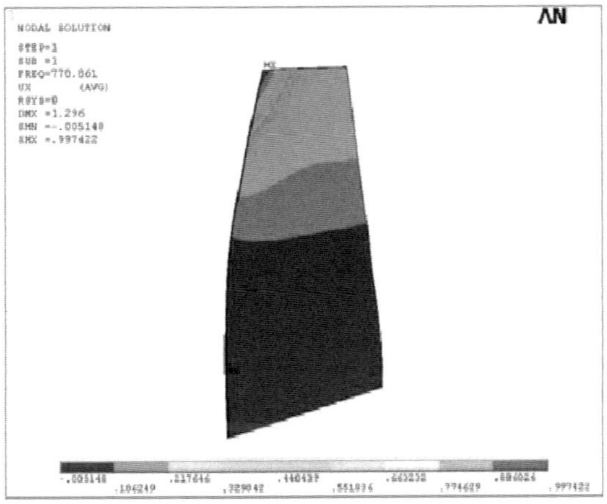


**Fig 17** First mode shape of carbon epoxy for (90/0/90) orientation with 7500Rpm

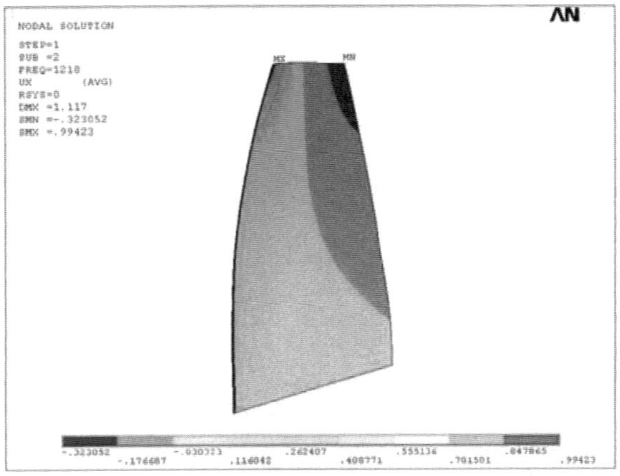


**Fig 18** Second mode shape of carbon epoxy for (90/0/90) orientation with 7500Rpm

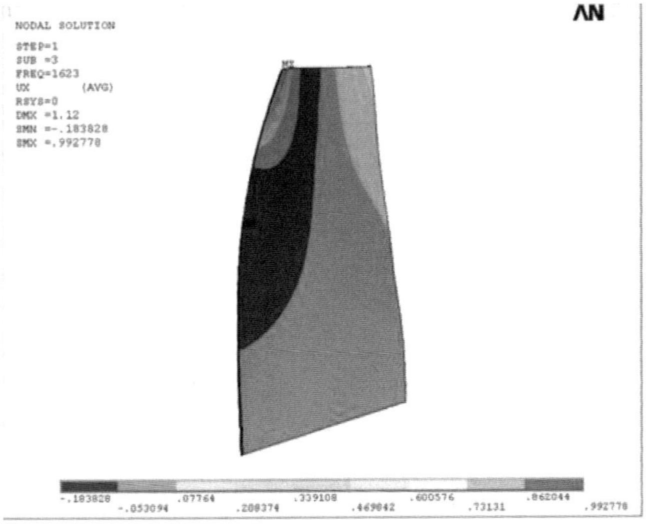


**Fig 19** Third mode shape of carbon epoxy for (90/0/90) orientation with 7500Rpm

The first mode shape of the carbon epoxy for three lamina orientation with 7500 rpm is similar to the first mode shape without pre-stress effect. The second mode shape with 7500 rpm is also similar to the second mode shape without pre-stress effect. The third mode shape with 7500 rpm varies from the third mode shape without pre-stress effect, such that the minimum displacement region from the middle portion of the leading edge to the middle portion of the tip of the blade

IOP Conf. Series: Materials Science and Engineering **247** (2017) 012007 doi:10.1088/1757-899X/247/1/012007

*3.2.4 MODE SHAPES OF THREE LAMINA CARBON EPOXY WITH 9000RPM FOR 90/0/90 ORIENTATION*

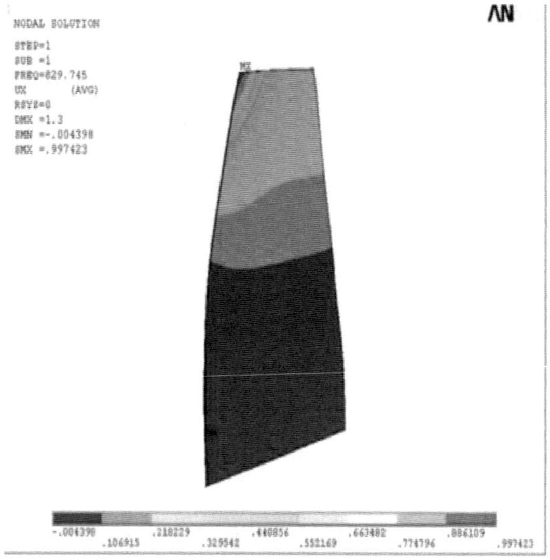


**Fig 20** First mode shape of carbon epoxy for (90/0/90) orientation with 9000Rpm

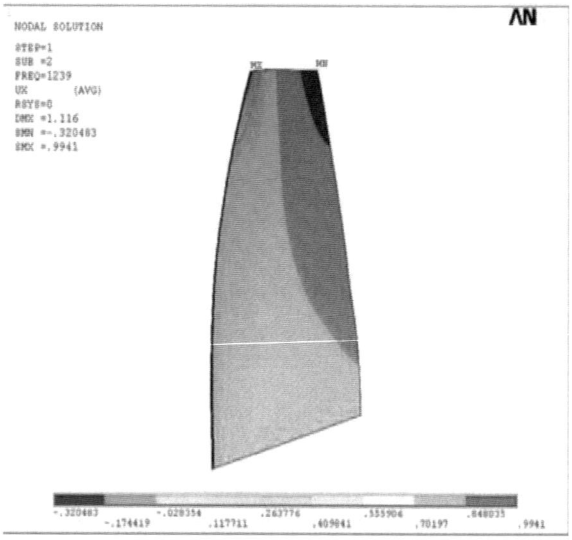


**Fig 21** Second mode shape of carbon epoxy for (90/0/90) orientation with 9000Rpm

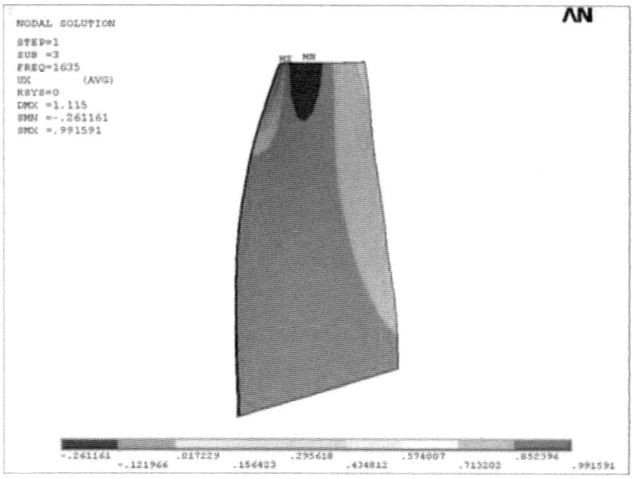


**Fig 22** Third mode shape of carbon epoxy for (90/0/90) orientation with 9000Rpm

IOP Conf. Series: Materials Science and Engineering **247** (2017) 012007 doi:10.1088/1757-899X/247/1/012007

The first mode shape with 9000 rpm is similar to the first mode shape without pre-stress effect. Second mode shape with 9000 rpm is also similar to that of second mode shape without pre-stress effect. The third mode shape with 9000 rpm varies with the third mode shape with 7500 rpm, such that the minimum displacement region completely shifted from the leading edge to the middle portion of the tip of the blade

### *3.2.5 CAMPBELL DIAGRAM FOR THREE LAMINA CARBON EPOXY FOR 90/0/90 ORIENTATION*

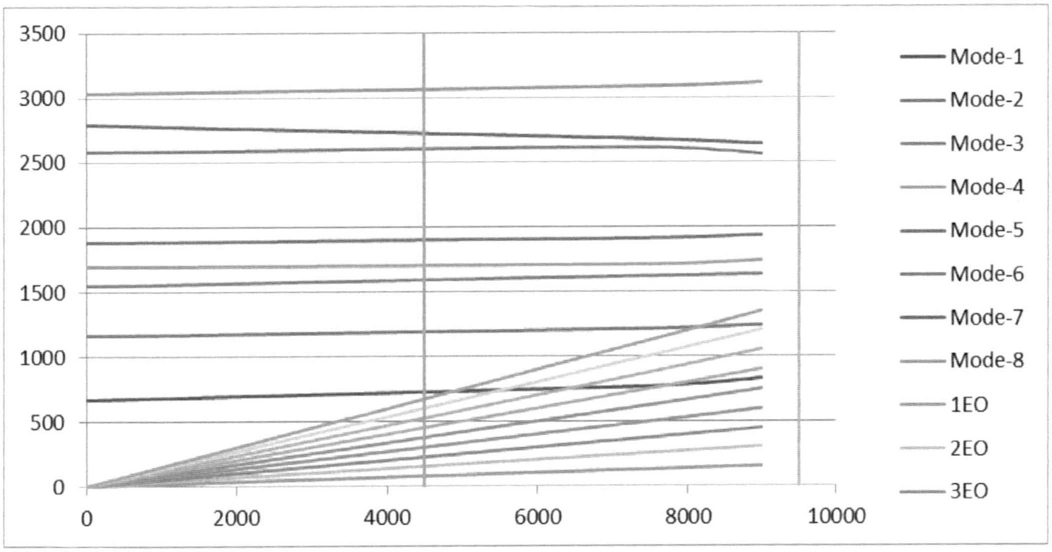


**FIG 23** Campbell diagram for 90/0/90 orientation

### *3.3 ANALYSIS OF BLADE PROFILE WITH FIVE LAMINA COMPOSITE MATERIAL*

Now extending the similar analysis for five lamina orientation as shown below, first all the four composite materials are analyzed for five lamina orientation. The Frequency values are tabulated as

IOP Conf. Series: Materials Science and Engineering **247** (2017) 012007 doi:10.1088/1757-899X/247/1/012007

**TABLE 7** Frequency values of composite materials with 5 lamina orientation

| Orientation | Carbon Frequency(Hz) | Graphite Frequency(Hz) | Aramid Frequency(Hz) | Glass Frequency(Hz) |
|---|---|---|---|---|
| **0/30/45/60/90** | 363.98 | 353.17 | 341.56 | 269.06 |
| | 913.18 | 893.95 | 839.41 | 674.63 |
| | 1184.1 | 1191.2 | 1056 | 851.01 |
| | 1646.6 | 1645.3 | 1473.5 | 1208 |
| **90/60/45/60/90** | 654.96 | 682.52 | 564.97 | 414.99 |
| | 1173 | 1180.4 | 1041.9 | 828.81 |
| | 1580.4 | 1617.8 | 1363.6 | 1076.2 |
| | 1749.2 | 1769.7 | 1546.4 | 1246.8 |

Similarly, from the above table it is clear that Carbon Epoxy and Aramid epoxy has good Dynamic characteristics. So, further study on lamina orientation is analyzed on Carbon Epoxy and Aramid epoxy for five different stacking sequence as shown,

**TABLE 8** Frequency values of carbon &aramid epoxy for 5 lamina orientation

| Orientation | Carbon Frequency(Hz) | Aramid Frequency(Hz) |
|---|---|---|
| 45/45/0/45/45 | 264.95 | 238.37 |
| | 727.62 | 665 |
| | 1021.6 | 943.67 |
| | 1472.7 | 1361.5 |
| 90/45/0/45/90 | 644.14 | 555.41 |
| | 1211.3 | 1067.5 |
| | 1577.4 | 1359.4 |
| | 1800.2 | 1587.5 |

### *3.3.1COMPOSITE BLADE WITH FIVE LAMINA ORIENTATION*

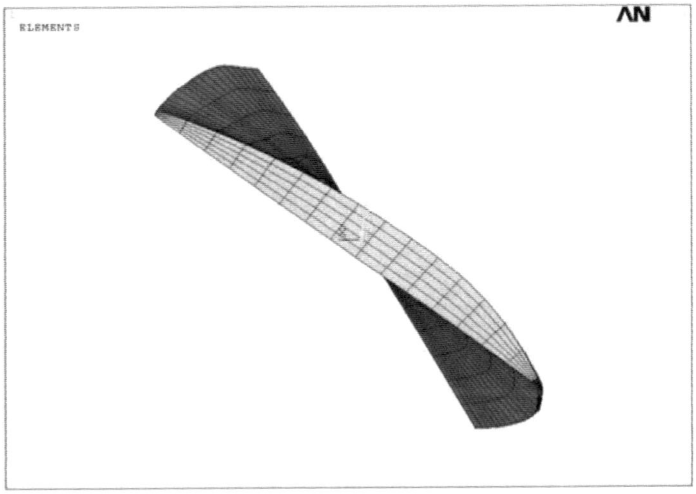


**FIG 24** Five lamina orientation on composite blade

Similar to three lamina orientation along the given trial blade profile, five laminae also prove that Carbon Epoxy has very good Dynamic Characteristics. So, Carbon Epoxy is analyzed for different Rpm condition. From the frequency values, it is understood that (90/45/0/45/90) ply orientation gives better values than other orientation. So, this particular orientation is used to extract Mode shapes and the Campbell diagram is generated.

**TABLE 9** Frequency values of 5 ply Carbon Epoxy with different RPM

| Orientation | 0 Rpm Frequency(Hz) | 7500 Rpm Frequency(Hz) | 9000 Rpm Frequency(Hz) |
|---|---|---|---|
| **45/45/0/45/45** | 264.95 | 468.72 | 561.19 |
| | 727.62 | 966.96 | 1084.6 |
| | 1021.6 | 1080 | 1125.5 |
| | 1472.7 | 1644 | 1667.5 |
| **90/45/0/45/90** | 644.14 | 747.84 | 809.39 |
| | 1211.3 | 1276.5 | 1312.1 |
| | 1577.4 | 1667.5 | 1725.3 |
| | 1800.2 | 1825.8 | 1833.5 |

### 3.3.2 *MODE SHAPES OF CARBON EPOXY WITH (90/45/0/45/90) PLY WITHOUT PRE-STRESS EFFECTS*

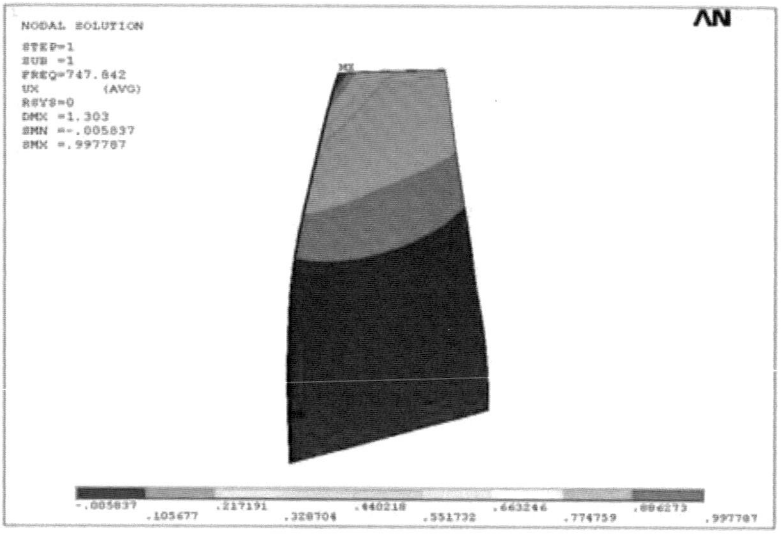


**FIG 25** First mode shape of Carbon Epoxy with (90/45/0/45/90) ply orientation

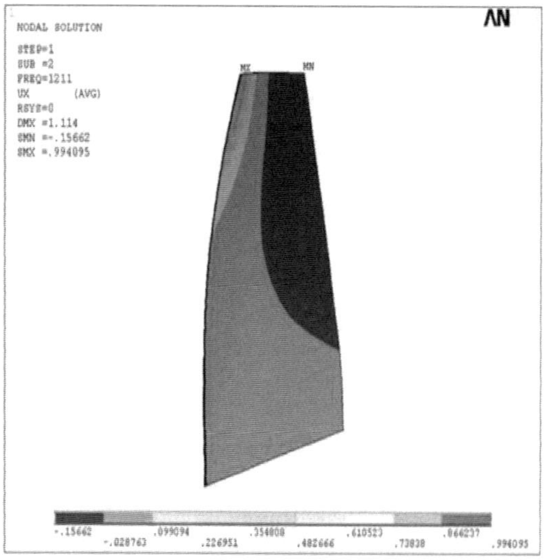


**Fig 26** Second mode shapes of Carbon Epoxy with (90/45/0/45/90) ply orientation

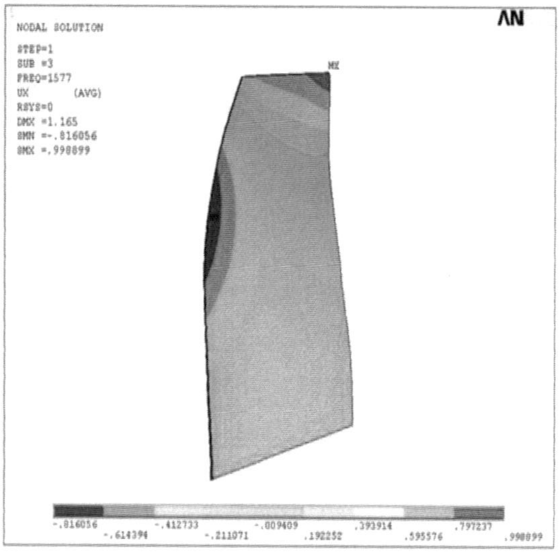


**Fig 27** Third mode shape of Carbon Epoxy with (90/45/0/45/90) ply orientation

In first mode shape for carbon epoxy for five lamina orientation without pre-stress effect, the maximum displacement will be at leading edge of the tip, while the minimum displacement will be at root of the blade.

In second mode shape for carbon epoxy for five lamina orientation without pre-stress effect, the maximum displacement will be at leading edge of the tip, while the minimum displacement will be at trailing edge of the blade. When compared to the mode shape of the carbon epoxy for three lamina orientation the minimum displacement region is increased. The third mode shape of the carbon epoxy for five lamina orientation completely differs from the mode shape of carbon epoxy with three lamina orientation.

*3.3.3 MODE SHAPES OF CARBON EPOXY WITH (90/45/0/45/90) PLY WITH 7500RPM*

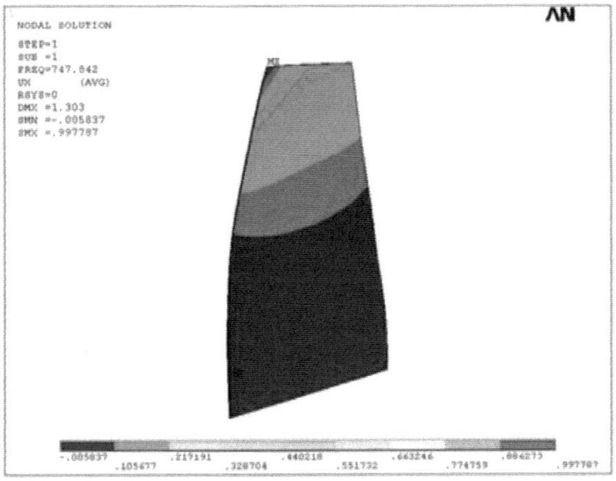


**Fig 28** First mode shape of Carbon Epoxy with (90/45/0/45/90) ply orientation

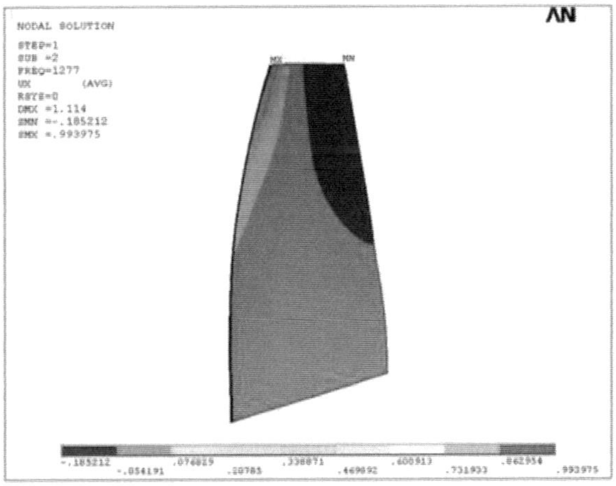


**Fig 29** Second mode shape of Carbon Epoxy w (90/45/0/45/90) ply orientation

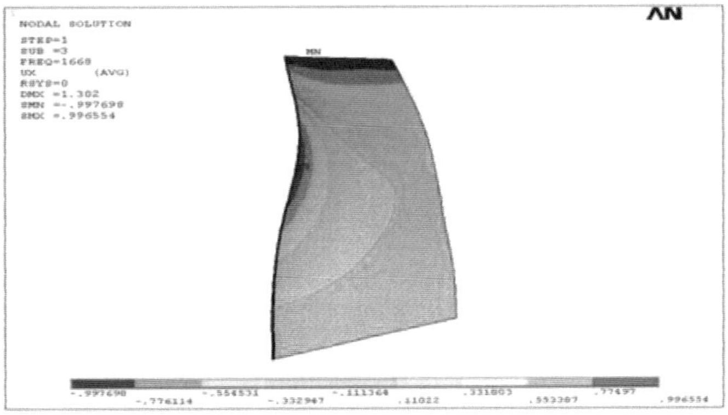


**Fig 30** Third mode shape of Carbon Epoxy with (90/45/0/45/90) ply orientation

The first mode shape with 7500 rpm is similar to the first mode shape without pre-stress effect. The second mode shape with 7500 rpm is similar to the second mode shape without pre-stress effect, the only difference is minimum displacement region is decreased slightly. The third mode shape with 7500 rpm completely varies from the third mode shape without pre-stress effect, such that the minimum displacement region is shifted from leading edge to tip of the blade

*3.3.4 MODE SHAPES OF CARBON EPOXY WITH (90/45/0/45/90) PLY WITH 9000RPM*

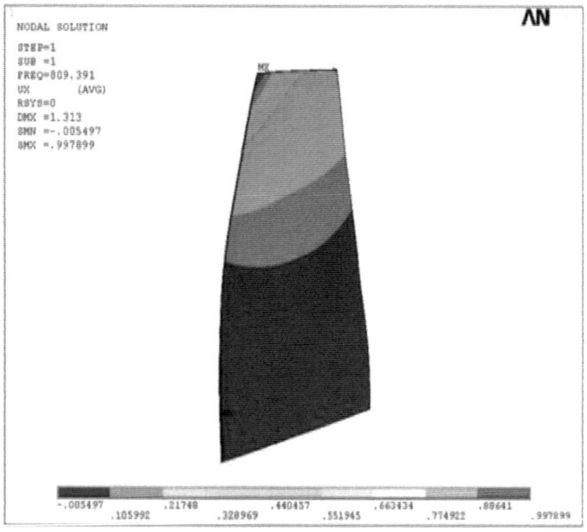


**Fig 31** First mode shape of Carbon Epoxy with (90/45/0/45/90) ply orientation

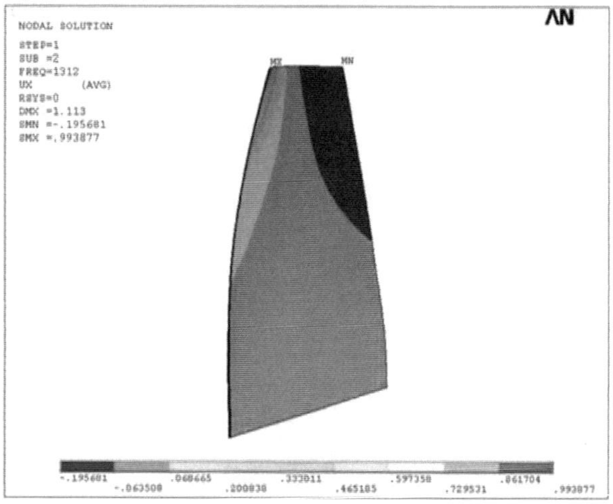


**Fig 32** Second mode shape of Carbon Epoxy with (90/45/0/45/90) ply orientation

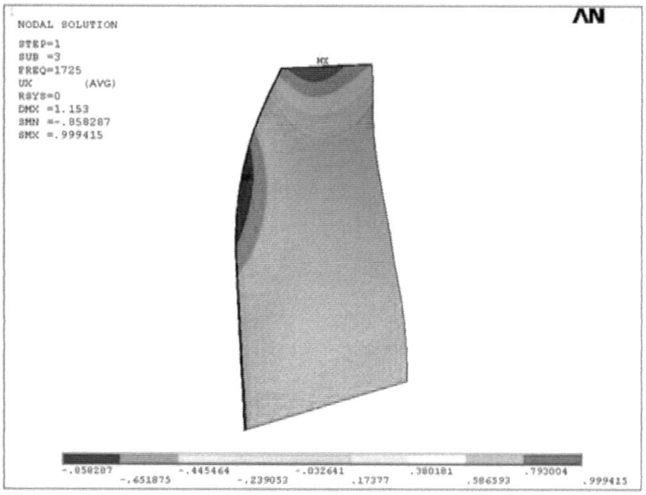


**Fig 33** Third mode shape of Carbon Epoxy with (90/45/0/45/90) ply orientation

The first mode shape with 9000 rpm is similar to the first mode shape without pre-stress effect. The second mode shape with 9000 rpm is similar to second mode shape with 7500 rpm, but the only difference is minimum displacement region is decreased. The third mode shape with 7500 rpm is similar to the third mode shape with 7500 rpm

*3.3.5 CAMPBELL DIAGRAM FOR CARBON EPOXY WITH (90/45/0/45/90) PLY*

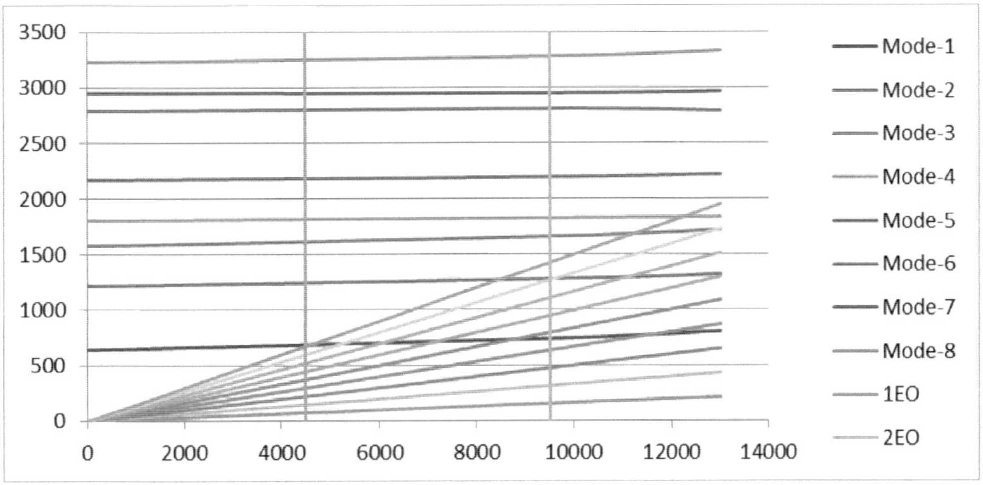


**FIG 34** CAMPBELL DIAGRAM FOR CARBON EPOXY WITH (90/45/0/45/90) PLY

*3.3.6 MASS COMPARISON OF ALL MATERIAL*

The drastic variation in the Mass of different Composite materials and Titanium is shown,

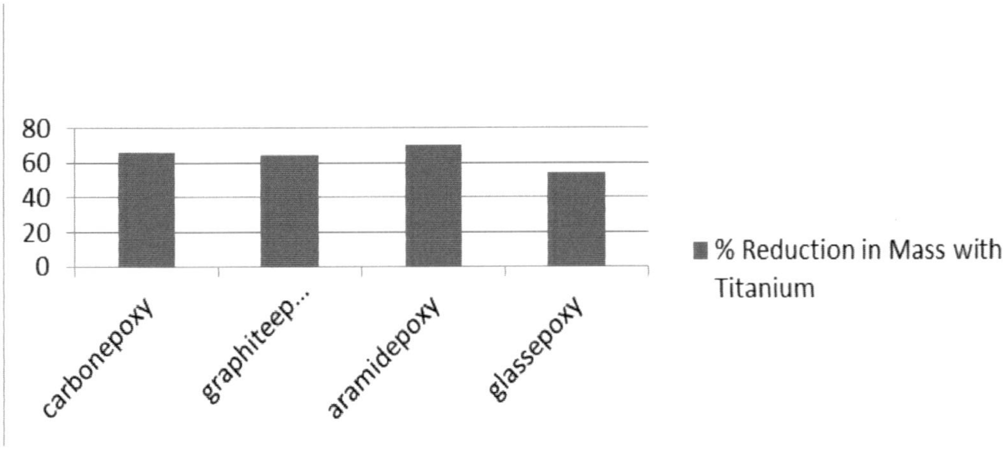


**FIG 35** Mass comparison between Composites & Titanium

It can be inferred from the chart that, the Composite material has less than 50% of the Mass of conventional Titanium material used. From the above chart it is seen that, Carbon Epoxy has only 41.4gms of mass while Titanium has 123.4gms each blade.

*3.3.7 FREQUENCY VALUES*

The First three modes frequency values of different Composite material are compared below,

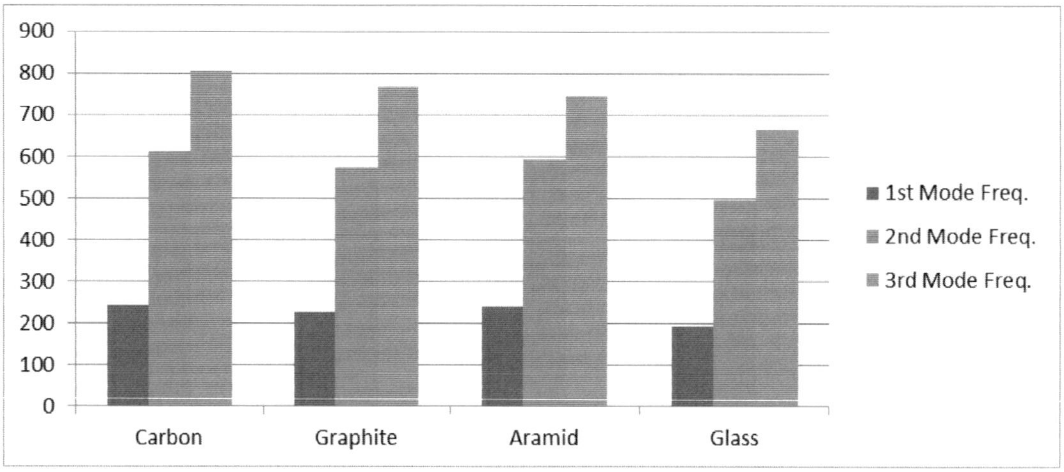


**FIG 36** Comparison of Frequency range among Composite materials

From the chart it is inferred that, Carbon Epoxy has higher Frequency range compared with other composite materials. Even all its mode frequency is higher than other material. Also from previous inference we understand that, Carbon Epoxy has good Dynamic Characteristics compared to other composite materials under study.

The Frequency values of various orientation of three ply Carbon Epoxy composite is shown below,

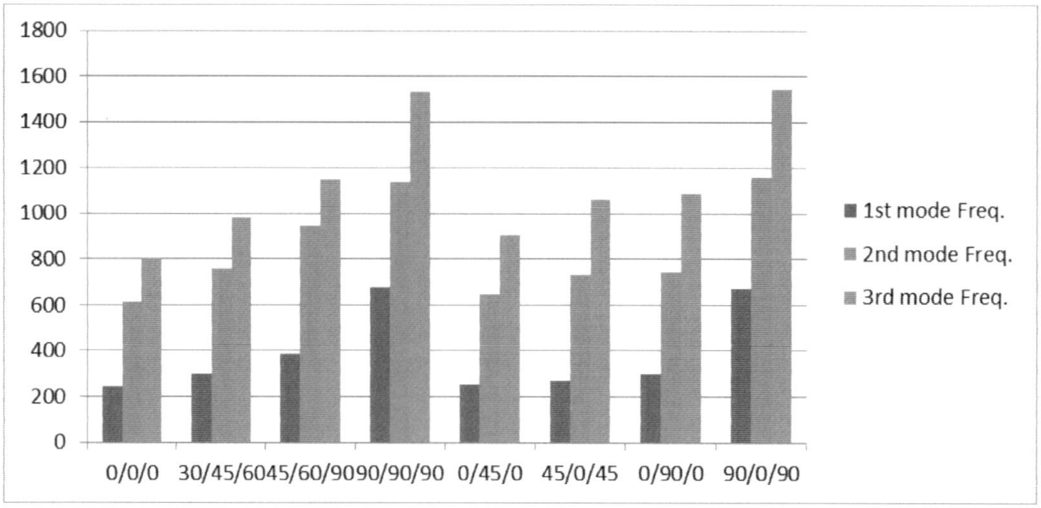


**FIG 37** Comparison of Frequency range of various orientation of 3ply Carbon Epoxy

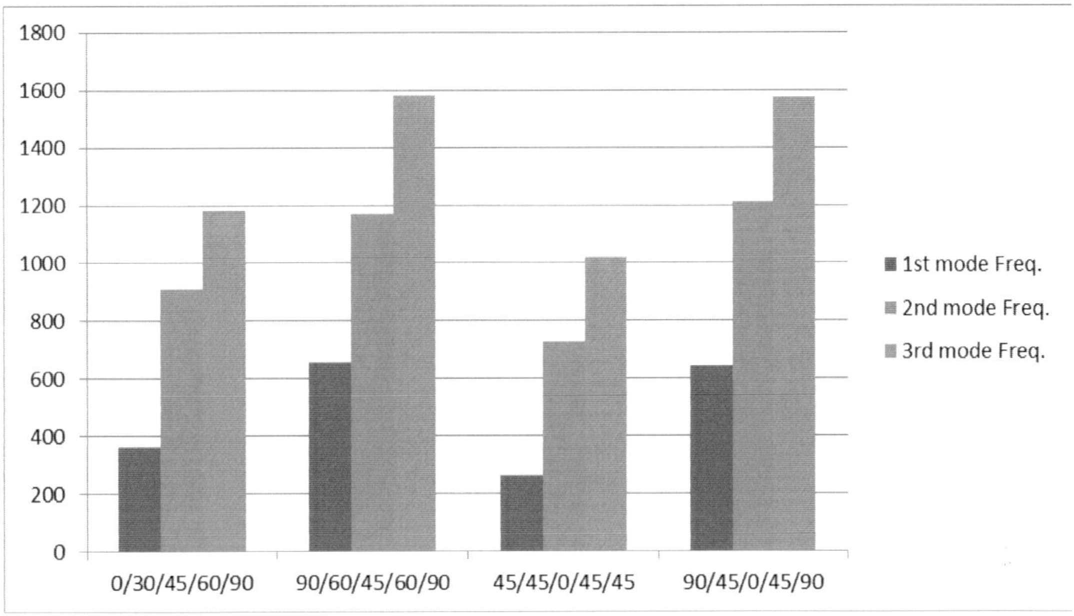


**FIG 38** Comparison of Frequency range of various orientation of 5ply Carbon Epoxy

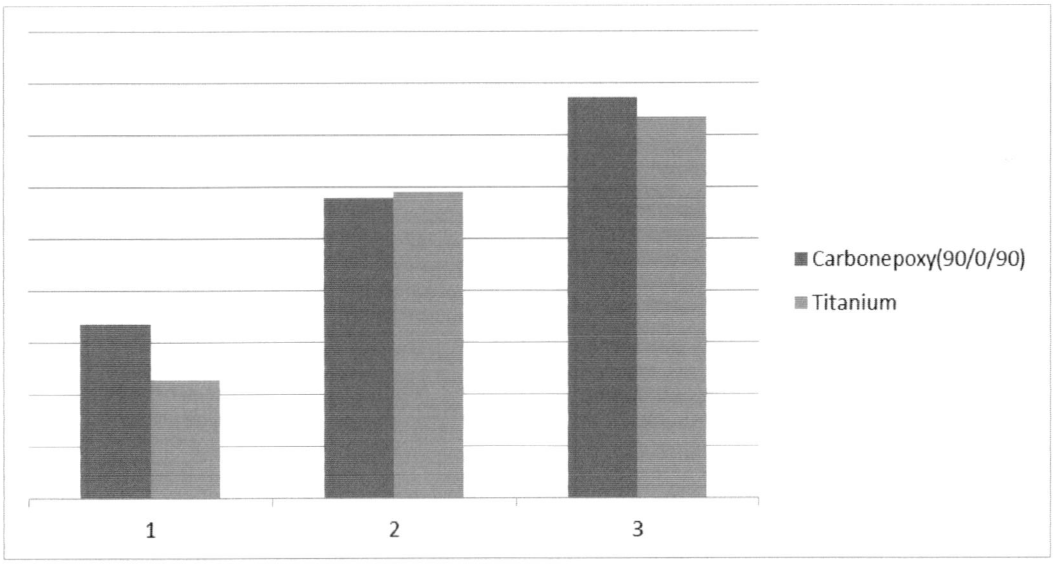


**FIG 39** comparison charts

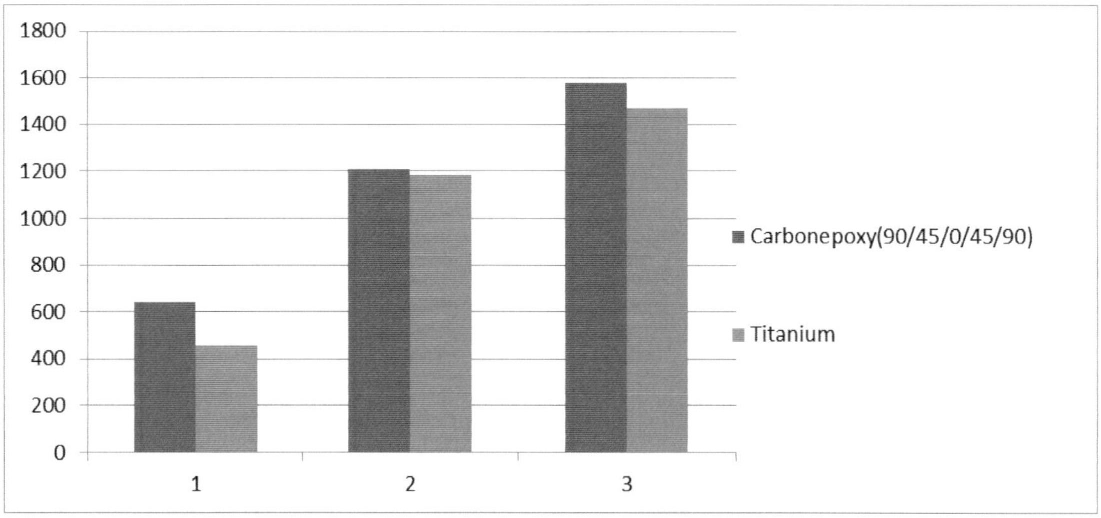


**FIG 40 comparison charts**

The charts above gives the comparison of Frequency values in (Hertz) between 3 ply, 5ply Carbon Epoxy composite material with Titanium alloy.

## 4. RESULT

- In this pilot work, the following issues are studied in detail to arrive at the feasibility of using composite laminates in place of metallic alloys such as titanium:
- Predict the natural frequencies and mode shapes of a typical titanium alloy blade configuration
- Calculate the weight of the blade configuration
- Predict the natural frequencies and mode shapes of the same blade configuration using the different composite materials identified for gas turbine usage.
- Calculate the weight at each case
- Analyze the structural dynamic characterization of each case and choose the feasible material
- Study the same material with different lamina orientations such as 0/0/0 and no of laminas along the thickness such as 3,5, etc.
- Choose the best possible configuration i.e. good improvement in natural frequency and mode shapes

The following table summarizes the results of good Composite materials in Dynamics point of view, from different composite material under study such as

- Carbon Epoxy
- Aramid epoxy
- Graphite epoxy
- Glass epoxy

For same orientation (90/0/90) for 3ply & (90/45/0/45/90) of 5ply composite material,

**TABLE 10** Comparison of satisfied 3ply composite material with titanium alloy

| MATERIAL | FREQUENCY (Hz) | % IMPROMENT IN 1s MODE FREQUENCY |
|---|---|---|
| **Carbon Epoxy (90/0/90)** | 672.86<br>1160.5<br>1544.9<br>1693.1 | 47.7 |
| **Aramid epoxy (90/0/90)** | 578.78<br>1000.4<br>1323.6<br>1476.4 | 27 |
| **Titanium alloy** | 455.42<br>1182.6<br>1469.5<br>2233.7 | Reference |

**TABLE 11** Comparison of satisfied 5ply composite material with Titanium alloy

| MATERIAL | FREQUENCY (Hz) | % IMPROMENT IN 1st MODE FREQUENCY |
|---|---|---|
| **Carbon Epoxy 90/45/0/45/90** | 644.14<br>1211.3<br>1577.4<br>1800.2 | 41.4 |
| **Aramid epoxy 90/45/0/45/90** | 555.41<br>1067.5<br>1359.4<br>1587.5 | 21.9 |
| **Titanium alloy** | 455.42<br>1182.6<br>1469.5<br>2233.7 | Reference |

IOP Conf. Series: Materials Science and Engineering **247** (2017) 012007 doi:10.1088/1757-899X/247/1/012007

From above analysis, it's evident that carbon epoxy composites with 3 ply &5ply shows a better structural dynamic characteristic.

From weight comparison between composites and titanium alloy, carbon epoxy shows significant weight reduction in %.

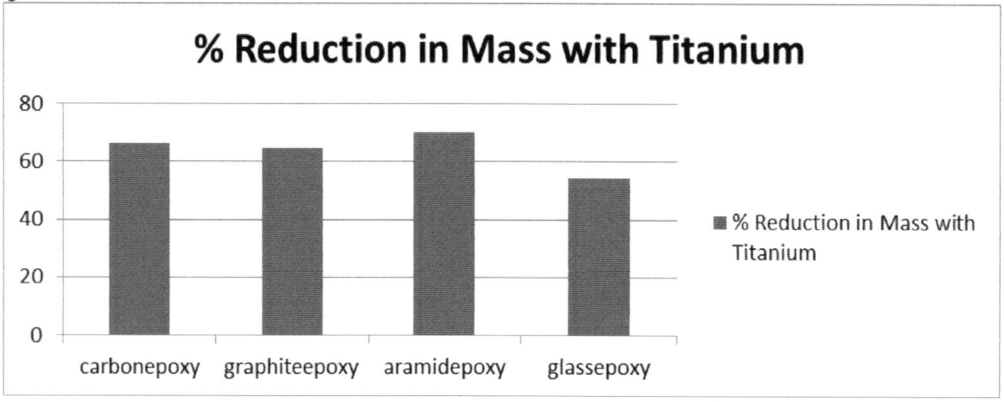


**FIG 41** mass comparison charts

And carbon epoxy is chosen for further analysis to study in detail and the results for various orientation are presented below.

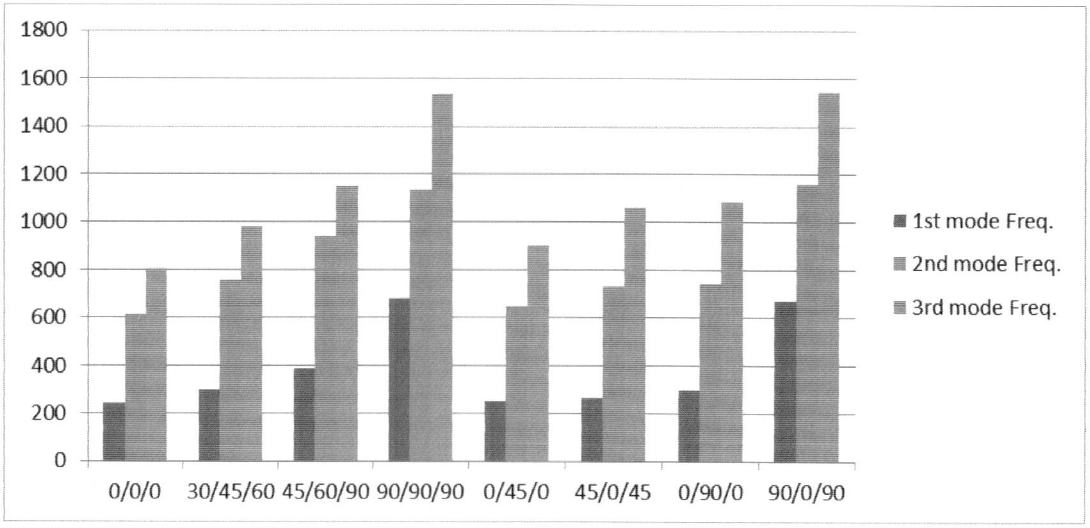


**FIG 42** frequency comparison charts

IOP Conf. Series: Materials Science and Engineering **247** (2017) 012007 doi:10.1088/1757-899X/247/1/012007

From above results it is evident that, the orientation (90/0/90) with no of lamina (Three) gives the natural frequency which has an improvement of (47.7%) over the Titanium alloy. And the orientation (90/45/0/45/90) with no of lamina (Five) gives the natural frequency which has an improvement of (41.4%) over the Titanium alloy. So, carbon epoxy with above orientation is considered for design.

## CONCLUSION

This result proves that it is possible to tailoring the mode shape, in case it is required by the designers. This mode shape tailoring is basically to avoid the high vibratory stresses at critical section in a blade. A wide variety of literature is available in this topic and the scope of the work is to only identify the possibly of change. A detailed work shall be carried out in future to study the implications of change in mode shapes and their relevance to gas turbine design.

## REFERENCE


[1] Dragan D. Kreculj, "Stress Analysis in an Unidirectional Carbon/Epoxy Composite Material," University of Belgrade Faculty of Mechanical Engineering.

[2] Gary D. Roberts and J. Michael Pereira, "Design and Testing of Braided Composite Fan Case Materials and Components," Glenn Research Center, Cleveland,Ohio

[3] T. Havar, "Analysis of Innovative Composite Aircraft Structures"

[4] Leihong Li "Structural Design of composite rotor blades with consideration of Manufacturability, durability, and manufacturing uncertainties". A Thesis Presented to The Academic Faculty .

[5] George marsh "Composites get in deep with new generation engine" , published in Reinforced plastics, Dec 2006.

[6] Galib H.Abumeri, Christos C.Chamus and Latif H. Kuguoglu "Composite fan blade design for advanced engine concepts," published in SAMPE 2003.

[7] Tery ford "Aerospace composites," published in Aircraft engineering andAerospace technology, Vol. 69

[8] R.Poul, D.Hanus,"Composite axial flow propulsor for small aircraft," Published in Acta polytechnia, Vol. 45

[9] J. Kubiak, G Urquiza, J.A. Rodriguez, G. González, I. Rosales, G. Castillo & J. Nebradt, Failure analysis of the 150MW gas turbine blades Engineering Failure Analysis, 16 (2009), pp. 1794–1804

[10] Carter T J, Common failures in gas turbine blades, Engineering Failure Analysis, 12(2005), pp. 237–247.

[11]. Yu-jiang Xie, Mao-cai Wang, Ge Zhang, Min Chang, Analysis of superalloy turbine blade tip cracking during service, Engineering Failure Analysis, 13 (2006), pp. 1429–1436.

[12] Ray A K., Failure mode of thermal barrier coatings for gas turbine blade under bending, International Journal of Turbo and Jet Engines, 17(2000),pp.1-24

# Design and Optimization of Aerospike nozzle using CFD

**Naveen Kumar K[1], Gopalsamy M[2], Daniel Antony[2], Krishnaraj R[3] and Chaparala B V Viswanadh[3]**

[1] Assistant Professor, Department of Aeronautical Engineering, Kumaraguru college of technology, Coimbatore, India.

[2] Assistant Professor, Department of Aerospace Engineering, Karunya University, Coimbatore, India.

[3] UG Scholar, Department of Aerospace Engineering, Karunya University, Coimbatore, India.

**E-mail:** kraj220@karunya.edu

**Abstract**. New rocket designs are being adopted to increase the performance of the current satellite launch vehicles (SLVs). But, the aerospike (or plug) nozzle concept that has been under development since the 1950s is yet to be utilized on a launch platform. Due to its ability to adjust the environment by altering the outer jet boundary, the aerospike nozzle delivers better performance compared to present day bell nozzle. An aerospike nozzle is designed for 20 bar pressure ratio. In order to improve the performance of the aerospike nozzle for various conditions, optimization of the nozzle was carried out for some important design parameters and their performances were studied for cold flow conditions. Initially a model of an aerospike nozzle is created for certain parameters, then the optimization process is carried out for the nozzle (Truncated model & Base bleed model). Optimized model is designed by the software GAMBIT and the flow behaviour is analysed by the Computational Fluid Dynamics (CFD) software called FLUENT. Comparison also takes place between the full length and the optimized models.

Keywords- Aerospike nozzle, Contour, Base bleed, altitude compensation

## 1. INTRODUCTION

The most popular altitude-compensating rocket nozzle to date is the aerospike nozzle, the origin of which dates back to Rocket dyne in the 1950s. This type of nozzle was designed to allow for better overall performance than conventional nozzle designs.

### *1.1. Nozzle*

A nozzle is a mechanical device of varying cross section which controls the direction and characteristics of the fluid (Air or Water) flowing through it. They are used in rocket engines to expand and accelerate the combustion gases, from burning propellants, so that the exhaust gases exit the nozzle at supersonic or hypersonic velocities.

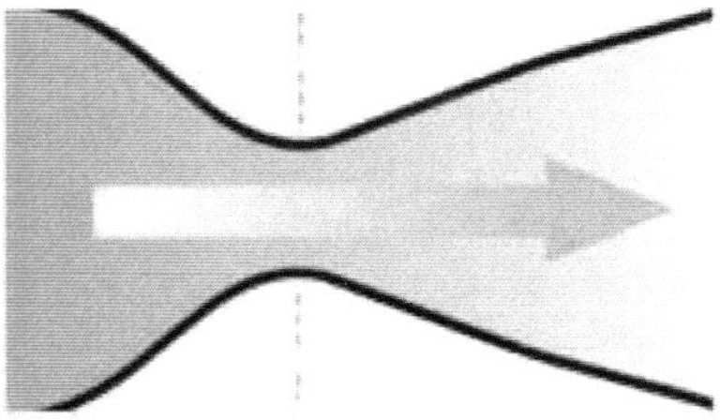

**Figure 1.** C-D Nozzle.

When the fluid flows through the nozzle it exits at a higher velocity than its inlet velocity. This phenomenon occurs due to conservation of mass which states that the rate of change of mass equals to the product of density, area and velocity. m = *A*V m = mass flow rate A = area of flow V = velocity of flow Solving this equation using differentiation, we get the equation,

$$\frac{dA}{A} = \frac{dV}{V}(1 - M^2)$$

*1.2. NEED FOR NEW DESIGN*

The revolution in aerospace propulsion was increased greatly during World War 2. Faster, bigger and more efficient aerospace vehicles were required which led to the birth of Space research organizations like NASA. Speaking about the future, advanced rocket propulsion systems will require exhaust nozzles that perform efficiently over a wide range of ambient operating conditions. Most nozzles either lack this altitude compensating effect or they are extremely difficult to manufacture. In present day, only bell nozzles are used for launch activities. But, these bell nozzles have a major drawback of decreasing in efficiency as altitude increases. This is due to a phenomenon causing loss of thrust in the nozzle at higher altitudes called as separation of the combustion gases.

For conventional bell nozzles, loss mechanisms fall into three categories:

- geometric or divergence loss,
- Viscous drag loss,
- Chemical kinetics loss.

Geometric loss results when a portion of the nozzle exit flow is directed away from the nozzle axis, resulting in a radial component of momentum. In an ideal nozzle, the exit flow is completely parallel to the nozzle axis and possesses uniform pressure and Mach number. By calculating the momentum of the actual nozzle exit flow and comparing it to the parallel, uniform flow condition, the geometric efficiency is determined. By careful shaping of the nozzle wall, relatively high geometric efficiency can be realized. A drag force, produced at the nozzle wall by the effects of a viscous high-speed flow, acts opposite to the direction of thrust, and therefore results in a decrease in nozzle efficiency. The drag force is obtained by calculation of the momentum deficit in the wall boundary layer. This viscous drag efficiency is defined as:

$$\eta_{drag} = 1 - \frac{\Delta C_{f(drag)}}{C_{f(ideal)}}$$

The third nozzle loss mechanism is due to finite-rate chemical kinetics. Ideally, the engine exhaust gas reaches chemical equilibrium at any point in the nozzle flow field, instantaneously adjusting to each new temperature and pressure condition. In real terms, however, the rapidly accelerating nozzle flow does not permit time for the gas to reach full chemical equilibrium.

$$\eta_{kin} = 1 - \frac{\Delta C_{f(ODK)}}{\Delta C_{f(ODE)}}$$

The overall nozzle efficiency is then given by the combined effects of geometric loss, viscous drag and chemical kinetics: $\eta_{kin}$= 0.99(approximate).

$$\eta_{overall} = \eta_{geo} * \eta_{kin} * (1 - \eta_{drag})$$

A long nozzle is needed to maximize the geometric efficiency; but simultaneously, nozzle drag is reduced if the nozzle is shortened. If chemical kinetics is an issue, then the acceleration of exhaust gases at the nozzle throat should be slowed by increasing the radius of curvature applied to the design of the throat region. The optimum nozzle contour is a design compromise that results in maximum overall nozzle efficiency. Nozzle contours can also be designed for reasons other than for maximum thrust. Contours can be tailored to yield certain desired pressures or pressure gradients to minimize flow separation at sea level. A nozzle contour designed to produce parallel, uniform exit flow, thereby yielding 100 % geometric nozzle efficiency, is called an ideal nozzle.

This ideal nozzle is extremely long and the high viscous drag and nozzle weight that results are unacceptable. Some design approaches consider truncating ideal nozzles keeping in mind the weight considerations. Most companies have a parabolic curve-fit program, generally used to approximate Rao optimum contours, which can also be used to generate desired nozzle wall pressures. For nozzles at higher altitudes, vacuum performance is the overriding factor relating to mission performance and high nozzle area ratio is therefore desirable. However, nozzle over-expansion at sea level does result in a thrust loss because the wall pressure near the nozzle exit is below ambient pressure. If the nozzles exit area could be reduced for launch and then gradually increased during ascent, overall mission performance would be improved. The ideal rocket engine would make use of a variable-geometry nozzle that adjusted contour, area ratio and length to match the varying altitude conditions encountered during ascent. This feature is referred to as Altitude Compensation.

### *1.3. Aerospike Nozzle*

The aerospike nozzle is a bell nozzle with its nozzle profile turned inside out. Flow of combustion gases is directed radially inward towards the nozzle axis. In the annular aerospike nozzle, flow issues from an annulus at a diameter located some radial distance from the nozzle axis. Flow is directed radially inward toward the nozzle axis. This concept is the opposite of a bell nozzle which expands the flow away from the axis along diverging nozzle walls. In an aerospike, the nozzle expansion process originates at a point on the outer edge of the annulus which is referred to as the "cowl-lip." In a standard bell nozzle, flow expansion continues regardless of what the ambient pressure is, and the flow can continue to over-expand until it separates from the nozzle walls. The linear aerospike, spike consists of a tapered wedge-shaped plate, with exhaust exiting on either side at the "thick" end.

#### *1.3.1. Working Principle of Aerospike nozzle:*

The function of a rocket nozzle is to direct all gases, generated in the combustion chamber of the engine and accelerated by the throat, out of the nozzle. The key feature of the aerospike engine is that, as the launch vehicle ascends during its trajectory, the decreasing ambient pressure allows the effective nozzle area ratio of the engine to increase. An aerospike nozzle is often referred to as an altitude-compensating nozzle, because of its specific design capability of maintaining aerodynamic efficiency as altitude increases and thus throughout the entire trajectory. At the outer cowl lip, the gas expands to the atmospheric pressure immediately, and then causes serious expansion waves propagating inward at an angle through the gas stream. At the location where the last expansion wave intercepts the spike, the gas pressure is equal to the atmospheric pressure. For the over expanded case, the spike changes the gas to be directed outward, and thus compression waves form and propagate outward at an angle and reflect off the jet boundary as expansion waves. This process then begins again. The aerospike features a series of small combustion chambers along the ramp that shoot hot gases along the ramp's outside surface to produce thrust in a spike-shaped plume, hence the name "aerospike."

IOP Conf. Series: Materials Science and Engineering **247** (2017) 012008 doi:10.1088/1757-899X/247/1/012008

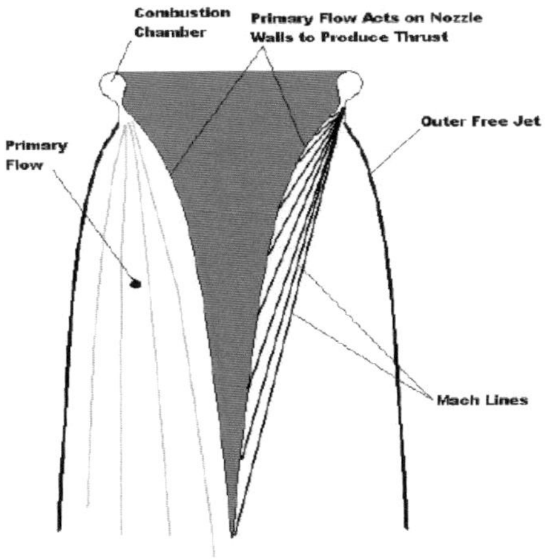


**Figure 2.** Model of aerospike with flow field

The ramp serves as the inner wall of the bell nozzle, while atmospheric pressure serves as the "invisible" outer wall. The combustion gases race along the inner wall (the ramp) and the outer wall (atmospheric pressure) to produce the thrust force.

*1.3.2. Flow around Aerospike Nozzle*

The main advantage to the annular aerospike nozzle design (both full length and truncated spike) is its altitude compensation ability below or at its design altitude. More specifically, the aerospike will not suffer from the same overexpansion losses a bell nozzle suffers and can operate near optimally, giving the highest possible performance at every altitude up to its design altitude. Above the design altitude, the aerospike behaves much like a conventional bell nozzle. Figure below shows the exhaust flow along an aerospike at low altitudes, design altitude, and high altitudes for a full spike and a truncated spike. Multiple expansion and compression, or shock, waves are evident in the flow in Figure these waves lead to losses in thrust. The outer flow boundary of the aerospike is the atmosphere itself. Unique to aerospike engines operating at their design altitude, engine geometry at the throat determines the expansion ratio of the aerospike nozzle and thus the corresponding engine performance.

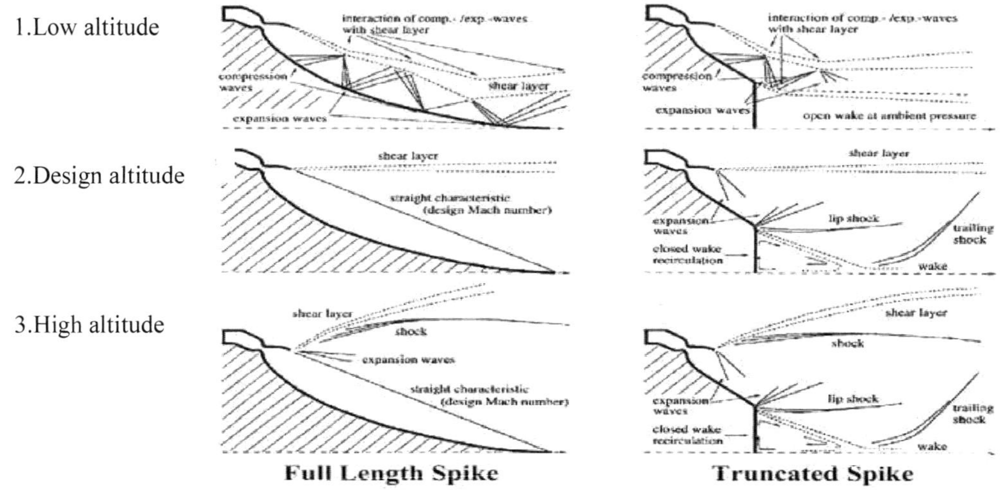


**Figure 3.** Exhaust Flow from a Full and Truncated Spike

At the design altitude of the nozzle, the exhaust flow at the chamber exit lip will follow a parallel path to the centreline to the exit plane. Therefore, the expansion ratio for a full-length spike at design altitude is equivalent to the chamber exit lip area divided by the throat area. As the ambient pressure decreases, the hot gas/ambient air boundary expands outward changing the pressure distribution along the spike; as a result, the expansion ratio increases. As the ambient pressure increases (low altitudes), the higher ambient pressure compresses the hot gas/ambient air boundary closer to the spike resulting in an expansion ratio decrease. The pressure distribution change along the spike and the location of the hot gas/ambient air boundary is automatic thus permitting altitude compensation up to the design altitude of the nozzle. Above the design altitude of the nozzle, the pressure distribution along the nozzle wall is constant. The expansion of the flow exiting the combustion chamber is governed by the Prandtl-Meyer turning angle at the throat.

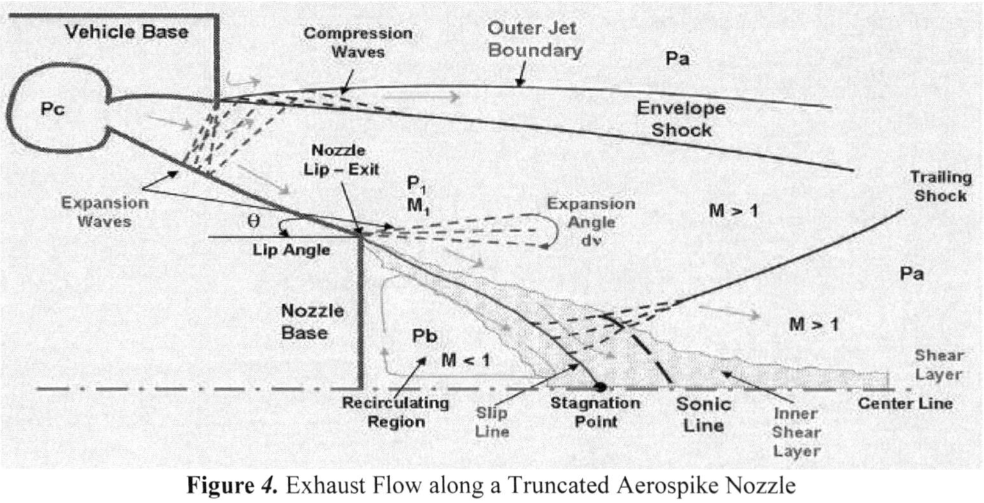


**Figure *4.*** Exhaust Flow along a Truncated Aerospike Nozzle

According to the aerospike nozzle numerical analyses the results of the altitude compensation capabilities of an aerospike up to the design altitude are undeniable. Furthermore, the aerospike performs worse at high altitudes compared to bell nozzles with equal expansion ratios (exit area divided by throat area); therefore, to get the benefit of the aerospike, the design pressure ratio and the expansion ratio should be chosen as high as possible. The design pressure ratio is the ratio of the chamber pressure to the ambient pressure; ambient pressure is based on the chosen design altitude. If the spike is truncated, the aerospike advantage at higher altitudes (orbit transfer missions) includes shortened nozzle length and lower mass as compared to an equivalent performance bell nozzle design for orbit transfer missions.

*1.3.3. Advantages and Disadvantages of Using an Aerospike Nozzle:*
The aerospike nozzle has 90% overall better performance than the conventional bell shaped nozzle. The efficiency at low altitudes is much higher because the atmospheric pressure restricts the expansion of the exhaust gas. A vehicle using an aerospike nozzle also saves 25-30% more fuel at low altitudes. At high altitudes, the aerospike nozzle is able to expand the engine exhaust to a larger effective nozzle area ratio. An aerospike nozzle with an expansion ratio of 200:1 to 300:1 can increase the thrust and specific impulse by five to six percent. Specific impulse is the total impulse per unit weight of propellant. As of now, the most widely used nozzle type is the bell-shaped nozzle. It has a high-angled expansion section, usually 20-50°, right behind the nozzle throat, which is then followed by a gradual reversal of nozzle contour slope so that the nozzle exit divergence angle is small, usually less than a ten-degree half angle. The drawback to using this type of nozzle is that its optimum design is for a specific altitude, and thus is better suited for multi-stage to orbit. The aerospike design is suitable for Single Stage to Orbit(SSTO)flight. There are rarely some disadvantages on using an aerospike nozzle.

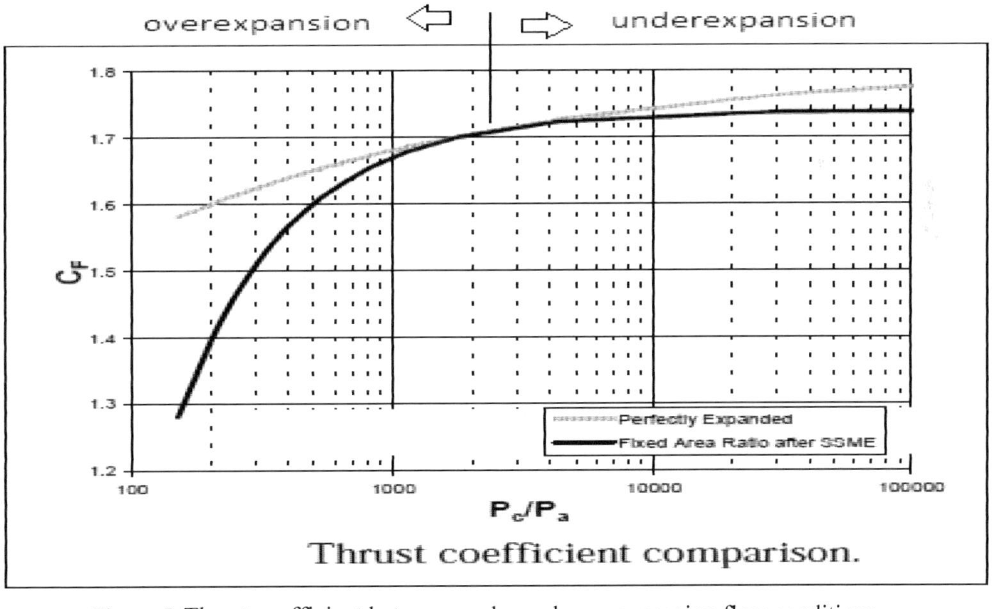


**Figure 5.** Thrust coefficient between under and over expansion flow conditions

The after body induces heat, and to cool means that the performance reduces along with the pressure against the nozzle. Another issue is weight, which as previously stated can be resolved through truncation. During flight a transonic and supersonic regime, generally between Mach 1 and 3, the slipstream effect reduces the aerospike nozzle's performance due to the external flow over the vehicle because the plume tends to draw in the air flow and thus alters the aerodynamics at the aft end of the vehicle. Finally, the performance is more difficult to evaluate because of the complex flow field and the turbulence involved.

IOP Conf. Series: Materials Science and Engineering **247** (2017) 012008 doi:10.1088/1757-899X/247/1/012008

### *1.4. Objective*

This thesis designs, analyses and optimizes the aerospike nozzle contours using CFD tool (FLUENT) and the prototype is fabricated from the design and tested in the ISRO (Mahendragiri) lab.

Considering that the rockets are designed for high-altitude applications (since most of its flight time is at high altitude) and the effects of density changes in the air will be negligible, the annular nozzle will probably be more appropriate for this application since the automatic altitude adjustment characteristics of the aerospike nozzle will not be necessary. However, in the interest of knowledge and the application of the data to future proposals, the aerospike design will be valuable and advantageous.

## 2. DESIGN METHODOLOGY

### *2.1. From Requirements to Preliminary Design:*

Since the cold flow is considered, the values of Specific Heat Ratio($\gamma$=1.4), gas constant of exhaust(R=287) and chamber temperature ($T_i$=300K) are taken of the values of ideal gas.

For every design there has to be minimum one parameter will be considered constant to determine the other design parameter values. Here, for the aerospike nozzle design, Mass Flow Rate(2Kg/S) and Nozzle Pressure Ratio(NPR=20) were considered as constant. Based on the value of Mass flow rate and ideal gas constant values, other basic design parameters were determined by 2D Isentropic Flow Equations.

#### *2.1.1. Plug Nozzle Contour Design:*

*Basic Design Methods:*

The design of spike contour of the aerospike nozzle is the most important step in the overall design of the nozzle, which varies according to the operating conditions and application. However, the design procedure, including the basic physics behind the spike design remains the same. The design of spike can be done using the following methods

i. Rao's method based on calculus of variation
ii. Simple approximate method
iii. B-Spline Method

Simple Approximation Method:

Here, Simple approximation method is chosen to design spike contour. The simple approximate method assumes a series of centred isentropic expansion waves occurring at cowl lip of the spike nozzle. Using this method, the annular spike contour for a given pressure ratio, area at throat, and ratio of specific heats can be determined.

A brief description of this is given below:

The expansion ratio is determined for the corresponding pressure ratio from the relation which specifies the variation between the two. Since the flow is assuming to be parallel to the nozzle exit the thruster angle is given by,

$$\theta_t = \vartheta(M_e)$$

Where $M_e$ is Mach number at exit and the Prandtl-Meyer function is given by

$$\vartheta = \sqrt{\frac{\gamma+1}{\gamma-1}}\tan^{-1}\sqrt{\frac{\gamma-1}{\gamma+1}(M^2-1)} - \tan^{-1}\sqrt{M^2-1}$$

The Throat area is,

$$A_t = \frac{\pi(r_e^2 - r_t^2)}{\cos\theta_t}$$

And exit area of the spike is,

$$A_e = \pi(r_e^2 - r_b^2)$$

Where $r_e$ denotes radius of cowl lip and $r_b$ denotes radius of base pressure, the chamber pressure F thrust and $C_f$ the thrust coefficient. Once the expansion ratio is calculated $r_e$ and $r_t$ can be determined for a fixed $A_t$.

The radial co-ordinate of spike nozzle is given by,

$$r^2 = r_e^2 - \frac{(r_e^2 - r_t^2)A\,sin(\mu + \theta)}{A_t\,sin(\mu)\,cos(\vartheta_t)}$$

The axial co-ordinate is given by,

$$x = \frac{r_e - r}{\tan(\mu + \theta)}$$

Where $\theta = \theta_t - \vartheta$ and Mach angle $\mu = sin^{-1}(\frac{1}{M})$.

The relationship for Mach number and area ratio is that given for isentropic expansion of a one-dimensional flow through varying area conduit. With the help of a FORTRAN program, the radial and the axial co-ordinates of the spike contour were found, thus obtaining the perfect shape of the spike satisfying the given problem statement.

### 2.2. *The FORTRAN Program*

Outline of External Expansion Plug Nozzle Design

**INPUT:**

- Estimate exit Mach number (which can be obtained from isentropic flow tables based on the expansion ratio of specific heats).
- Expansion ratio.
- Number of contour points
- Gas constant
- Exit temperature
- Atmosphere pressure ratio
- Constant of proportionality in Newton's second law
- Ratio of specific heats (constant or variables)

**OUTPUT:**

- Angle between plug axis and sonic line
- Width of throat gap
- Optimum thrust coefficient

IOP Conf. Series: Materials Science and Engineering **247** (2017) 012008 doi:10.1088/1757-899X/247/1/012008

- Mach number distribution
- Co-ordinates of plug contour
- Pressure ratio at each point
- Cumulative vacuum thrust coefficient
- Cumulative specific impulse
- Cumulative vacuum specific impulse

*2.2.1. Flow chart of External Expansion Plug Nozzle Design:*

C.C. Lee (1963) created and verified the FORTRAN program for external plug nozzle based on the simple approximation method which is used here.

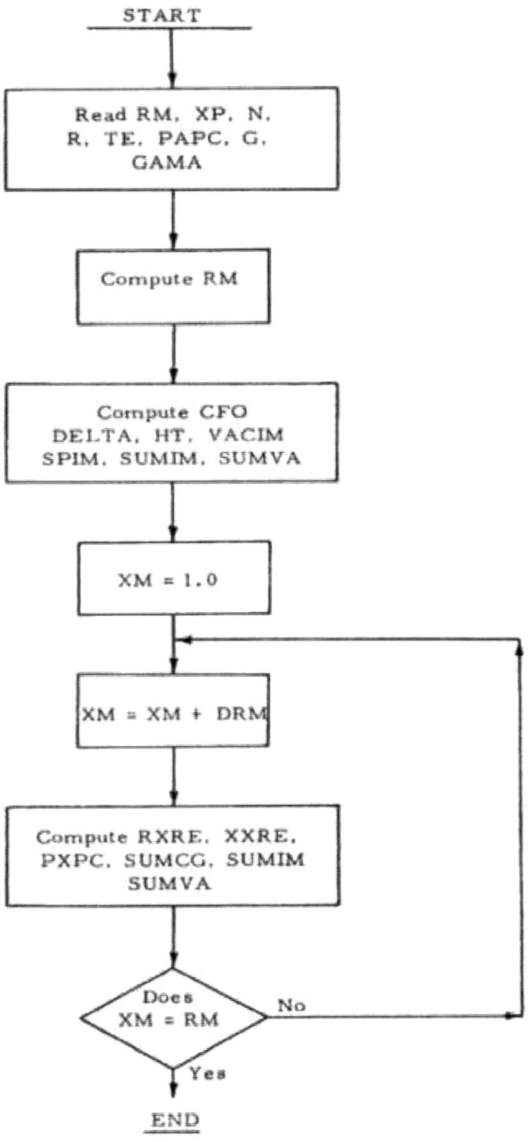


**Figure 6.** FORTRAN Flow chart of External Expansion Plug Nozzle Design

*2.3. Governing equations*

To simulate gas flow in the primary nozzle and on the outside of the plug, the compressible Reynolds-averaged Navier-Stokes (RANS) equations are employed. The following assumptions are made in the derivation on the appropriate form of the equations: compressible flow, Newtonian fluid, negligible gravity effects, no slip condition, 2-dimensionality, no chemical reactions, and thermal equilibrium.
The equations are solved using an implicit, coupled, three-dimensional finite volume method. A second-order accurate, upwind discretization scheme is used for the convective term and multigrid scheme is used to accelerate the rate of convergence. Fluent 6.3 is used to perform these calculations. Considering that the velocity of the flow is in supersonic, inviscid turbulence models are employed.

The ambient pressure does not affect the flow in the primary nozzle. Therefore, to simplify the computation, in this study the flow fields over the plug wall are considered as the environment and the atmospheric boundary condition is given to it throughout the calculation.

## 3. CFD SETUP AND SOLUTION

The designed aerospike nozzle was analyzed using commercially available CFD software Fluent 6.3. The modeling of the domain was made using Gambit 2.3 software. It uses Finite volume technique to solve the Navier Stokes equation. Proper meshing was done to ensure the accuracy of the solution.

### *3.1. Design Parameters:*

The input parameters for the design of the annular aerospike nozzle are,

- Exit Area ratio (Ae/At) = 2.8998
- Exit diameter Re = 18mm
- Chamber Pressure = 20 bar
- Exit Pressure = 1 bar
- Mass Flow Rate = 1.988Kg/Sec

From the given interval of design parameters, some of the values that are used for the design are,

- Length of the nozzle = 60mm
- Thrust = $1.268E^3$N
- Coefficient of thrust ($C_F$) = 1.446
- Mach No at the end of the ramp = 2.601

### *3.2. COMPUTATIONAL DOMAIN CREATION (GAMBIT):*

The computational domain of the annular Aerospike nozzle modelled in Gambit. Boundary conditions are imposed in this Computational domain. The distance from the tip of the plug to the end of the imaginary boundary is 10 and 5 times of the nozzle outer diameter with respect to X and Y axis. The nozzle geometry and flow is axisymmetric, so upper half of the nozzle only designed for flow analysis and thus reduce the computational time. The modelled nozzle and the nozzle with the computational domain is shown below,

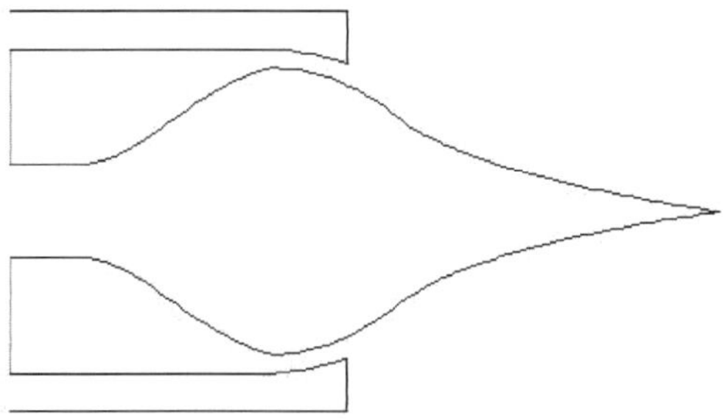

**Figure 7.** Designed aerospike nozzle in GAMBIT.

**Figure 8.** Aerospike nozzle design with far field boundary

### *3.2.1. Grid Generation:*

Quad-map meshing element is used to mesh the entire domain. The intensity of the grid is more near the wall and the regions where the gradients are high. The grid independency study was carried out and it was found that a minimum of 27000 nodes were needed. The total numbers of nodes on the domain are 27998 and cells are 27396.

**Figure 9.** Aerospike nozzle design after mesh

IOP Conf. Series: Materials Science and Engineering **247** (2017) 012008 doi:10.1088/1757-899X/247/1/012008

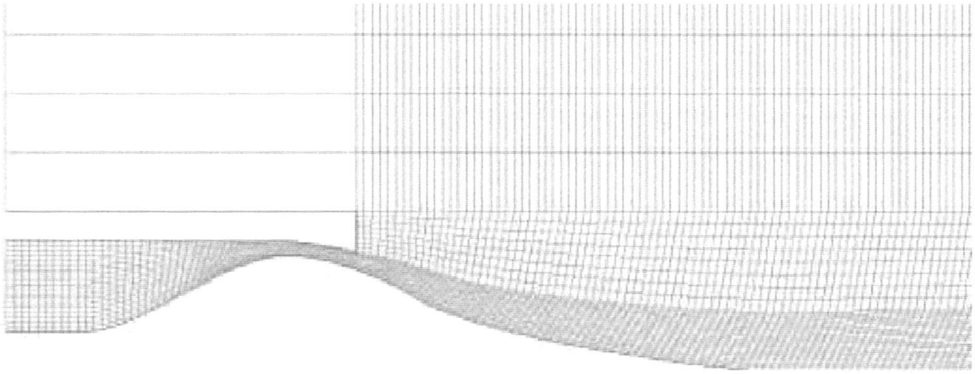

**Figure 10.** High density grids on the areas where high flow gradient occurs.

*3.3. FLOW PARAMETERS AND TURBULANCE MODEL (FLUENT):*

After modelling and meshing the nozzle in GAMBIT 2.3, CFD analysis made by using FLUENT 6.3 to solve the Navier-Stokes Equations. Air is used as the fluid medium. The material property of the gas is ideal gas. Since, the Mach No > 0.3, The ambient conditions are supposed at the far field Boundary Conditions.

FLUENT Boundary Conditions used for simulations are given below,

| | | |
|---|---|---|
| i. | Solver | = Density based |
| ii. | Space | = Axisymmetric |
| iii. | Viscous model | = Spalart – Allmaras, RNG k-ε |
| iv. | Operating Pressure | = 0 Pa |

*Boundary conditions*

*Pressure Inlet*

| | | |
|---|---|---|
| i. | Gauge total pressure | = 8 bar |
| ii. | Total temperature | = 300 K |
| iii. | Direction Specification | =Normal to boundary |

*Pressure Outlet*

| | | |
|---|---|---|
| i. | Gauge total pressure | = 1 bar |
| ii. | Backflow total temperature | =300K |

*Far Field Conditions*

| | | |
|---|---|---|
| i. | Gauge pressure | = 1 bar |
| ii. | Mach Number | = 0.4 |
| iii. | Temperature | = 300K |

*Solution Control*

| | | |
|---|---|---|
| i. | Discretization flow | = Second order upwind |
| ii. | Solver parameters | = Courant No (3) |

*3.3.1. Contours of Full Length Nozzle:*
*Mach Contour*

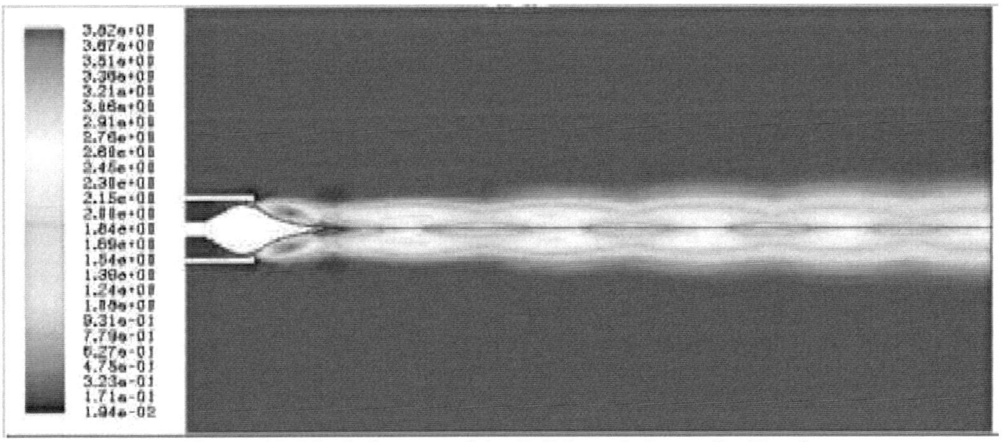


**Figure 11.** Mach Contour of the Full Length Aerospike Nozzle

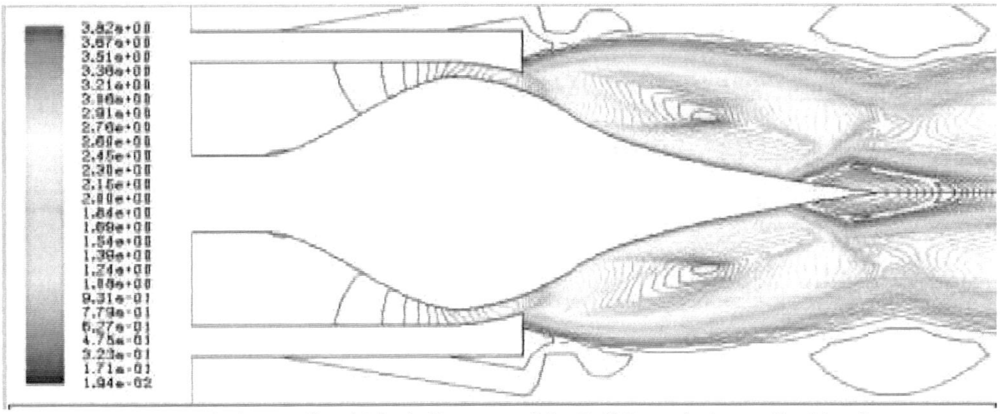


**Figure 12.** Streamlined Mach Contour of the Full Length Aerospike Nozzle

IOP Conf. Series: Materials Science and Engineering **247** (2017) 012008 doi:10.1088/1757-899X/247/1/012008

### *3.4. Optimization studies*

Need of Optimization for Aerospike Nozzle:

The wake formation at the end of the plug is studied. The formation of wake at the surface of the nozzle may affect the thrust production of the nozzle. To overcome this problem, the various optimized models of the same aerospike nozzle is designed. Various models were considered & compared. The Performance of Full length & optimized models were compared.

Nozzle design parameters considered for optimization:

i) Effect of truncation:

10%,20%,40%,60% & the point on the axial length where the Mach No starts to decrease considered for truncation.

ii) Effect of base bleed:

Base bleed systems combined with the above mentioned truncated models (20%, 40%, 60% &the point where the Mach No maximum) are considered for optimization.

iii) Effect of thruster Area Ratio:

Current nozzle design has the thruster area ratio(TAR) of 1.2. In order to obtain better performance, designs having Thruster area ratio of 1.3,1.4 and 1.5 are designed and their analyzed results were taken for comparison.

#### *3.4.1. Designs of Optimized models:*

1. Truncated model:

In the effect of truncation criteria, the truncated models of 10%,20%,40%,60% & the point on the axial length were the Mach Number starts to decrease are designed and analysed for the same flow conditions specified above.

The table below shows the highest Mach number attained on truncated models:

**Table 1.** Optimized Truncated models Mach Number values.

| TRUNCATED MODEL | MACH NO ATTAINED IN S-A TURBULANCE MODEL | MACH NO ATTAINED IN K-ε TURBULANCE MODEL |
|---|---|---|
| Full Spike | 3.82 | ------ |
| 10% | 3.82 | ------ |
| 20% | 3.56 | 3.57 |
| **40%** | **3.54** | **4.47** |

| | | |
|---|---|---|
| 60% | 3.51 | 4.46 |
| Truncated at Max Mach number attained point on the wall(17.5mm) | 3.52 | 4.39 |

2. Base bleed model:

In the truncated model, the formation of wake on the base of the nozzle creates pressure drops which cause reduction in its thrust producing capability. So, by providing additional flow through the nozzle to its base will reduce the wake formation. For the additional flow a hole has created in the nozzle's solid part from the pressure inlet portion to the base of the nozzle.

The only drawback of this model is it requires excess amount of mass flow rate for base bleed operation.

The table below shows the highest Mach no attained on truncated models:

**Table 2.** Optimized Truncated Base bleed models Mach Number values.

| TRUNCATED MODEL | BASE DIA (mm) | BLEED HOLE DIA (mm) | MASS FLOW RATE (Kg/S) | EXCESS % OF MASS FLOW RATE FOR BASE BLEED |
|---|---|---|---|---|
| 20% | 3.918 | 2 | $1.2699E^{-3}$ | .6356 |
| **40%** | **9.176** | **4.588** | $\mathbf{6.6833E^{-3}}$ | **3.345** |
| 60% | 17.026 | 8.512 | $23.004E^{-3}$ | 11.5135 |
| @ Max Mach no attained pt. | 23.462 | 8.512 | $23.004E^{-3}$ | 11.5135 |

3. Effect of Thruster Area Ratio:

Current model has the thruster area ratio of 1.2. By increasing the thruster area ratio, nozzle can produce more thrust. When the Thruster Area Ratio changes, mass flow rate, thrust and Mach No at the exit of the thruster also changes. These changing values based on the thruster area ratio are tabulated below,

IOP Conf. Series: Materials Science and Engineering **247** (2017) 012008 doi:10.1088/1757-899X/247/1/012008

**Table 3.** Optimized Various Aspect ratio models Mach Number Values

| MODEL BASED ON THRUSTER AREA RATIO | MACH NO AT THE END OF THE THRUSTER | THRUST (KN) | MASS FLOW RATE (Kg/S) |
|---|---|---|---|
| 1.1 | 1.372 | 1.403 | 2.02 |
| **1.2** | **1.534** | **1.268** | **1.988** |
| 1.3 | 1.659 | 1.167 | 1.949 |
| 1.4 | 1.763 | 1.089 | 1.915 |
| 1.5 | 1.854 | 1.027 | 1.887 |

*3.5. Selection of Optimized model*

The three aspects shown below are considered before concluding the optimized model.

- Thrust Production.
- Weight reduction.
- Low Mass Flow Rate increment due to optimization.

Based on the points mentioned above and with reference to the Tables which has analysed results of the optimized models, the 40% truncated aerospike nozzle and 40% truncated base bleed is chosen to compare with the Full-length aerospike nozzle.

*3.5.1. 40% Truncated Model:*

For the 40% length external aerospike nozzles, the outlet boundary of the nozzle was created with one endpoint having the coordinates of the last point on the full-length nozzle far field contour and the other endpoint having the x-coordinate of the last point on the 100% length contour and an y-coordinate equal to the y-coordinate of the expansion point.

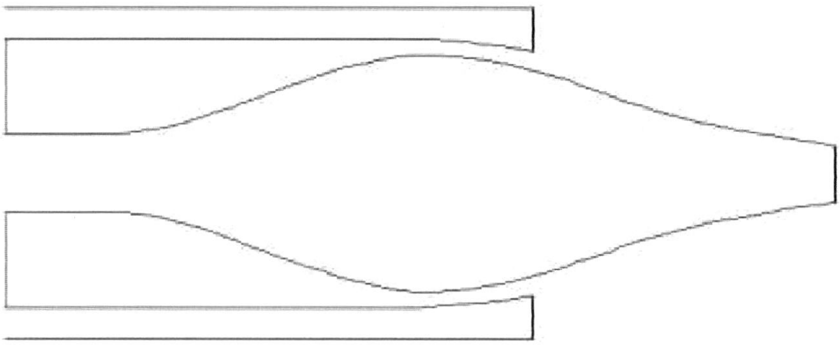

**Figure 13.** Designed 40% truncated aerospike nozzle in GAMBIT

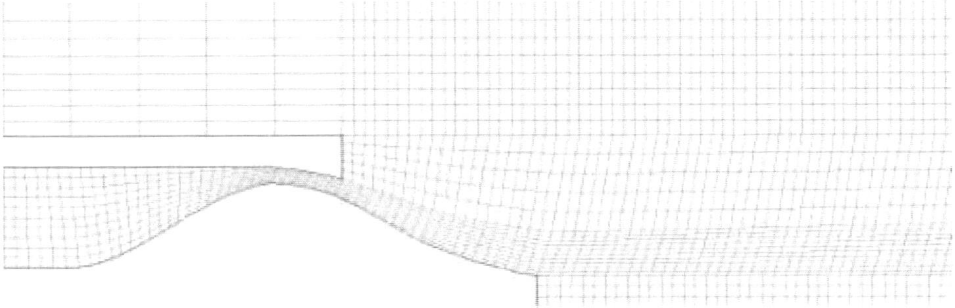

**Figure 14.** High density grids on the areas where high flow gradient occurs

*Contours of 40% Truncated Nozzle:*
*Mach Contour*

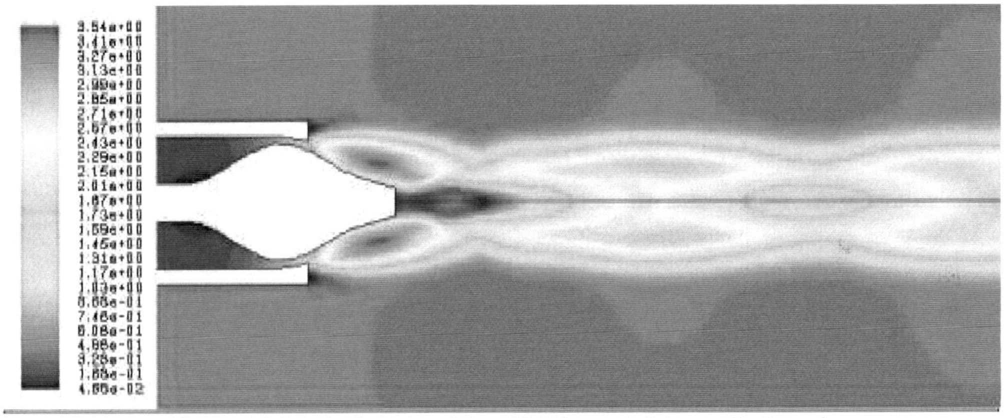


**Figure 15.** Mach Contour of the 40% Truncated Aerospike Nozzle

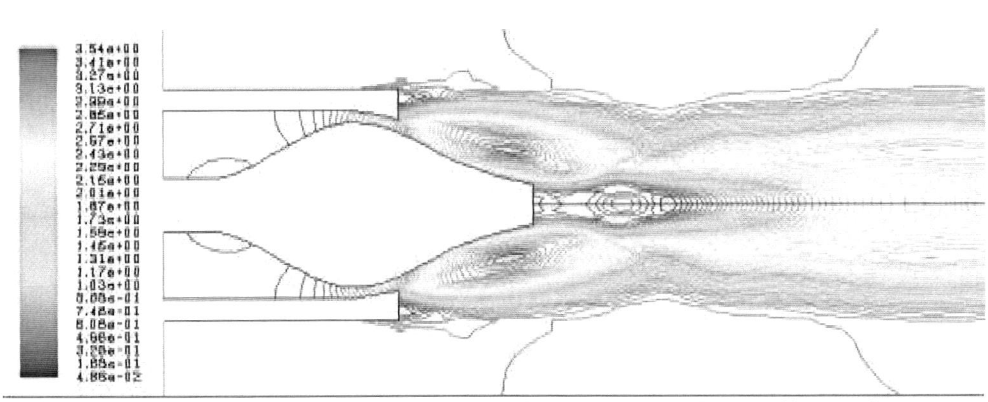


**Figure 16.** Streamlined Mach Contour of the 40% Truncated Aerospike Nozzle.

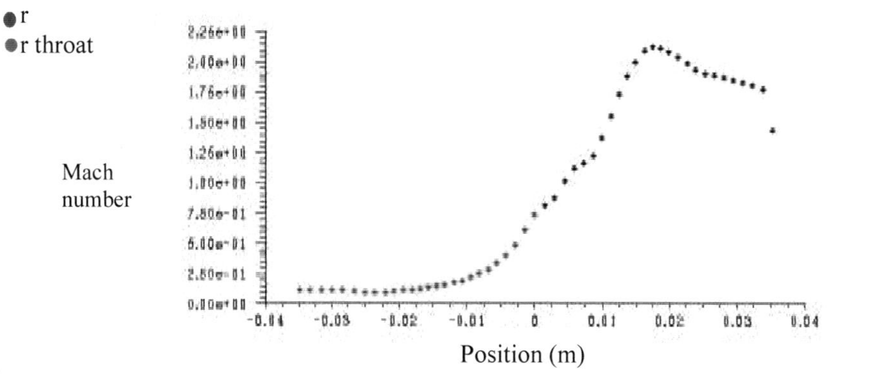


**Figure 17.** Mach Number Vs Ramp wall plot of the 40% Truncated Aerospike Nozzle

*3.5.2. 40% Truncated Base Bleed Model:*

Since the 40% truncated model is taken as the perfect optimized model, the base bleed model also designed on the 40% truncated model.

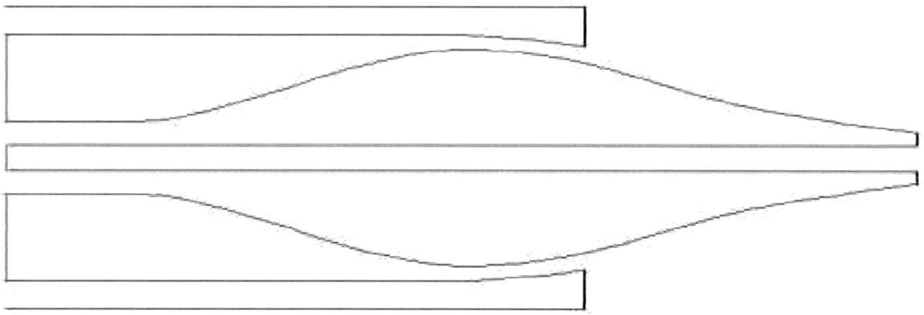

**Figure 18.** Designed40% truncated base bleed aerospike nozzle in GAMBIT.

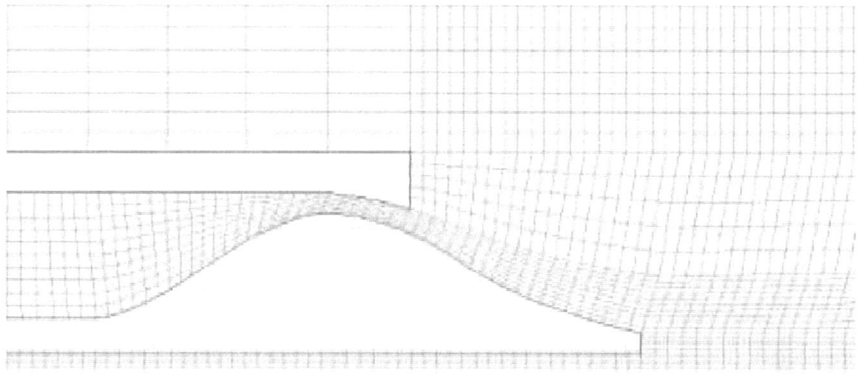

**Figure 19.** High density grids on the areas where high flow gradient occurs

*Contours of 40% Truncated base bleed model:*

*Mach Contour*

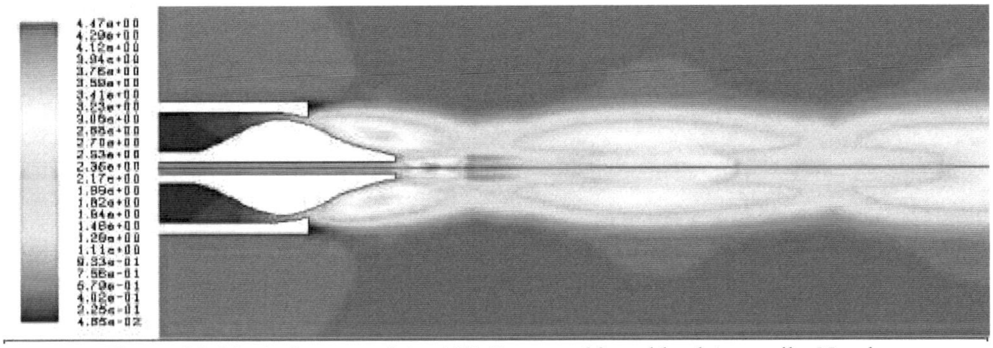


**Figure 20.** Mach Contour of the 40% Truncated base bleed Aerospike Nozzle

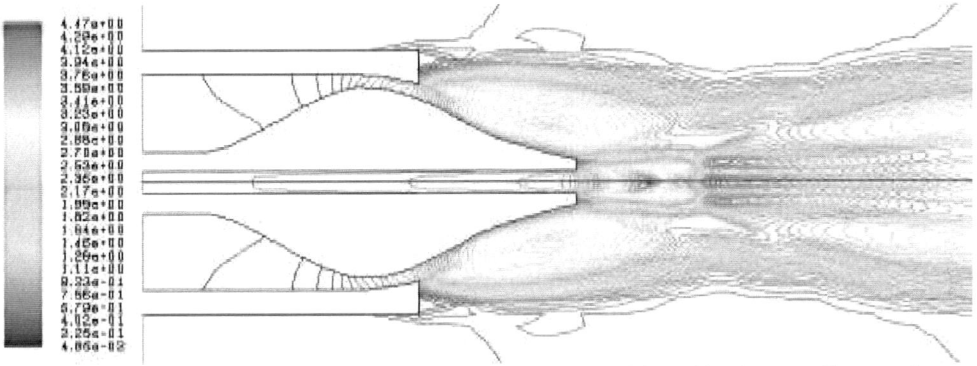


**Figure 21.** Streamlined Mach Contour of the 40% Truncated base bleed Aerospike Nozzle.

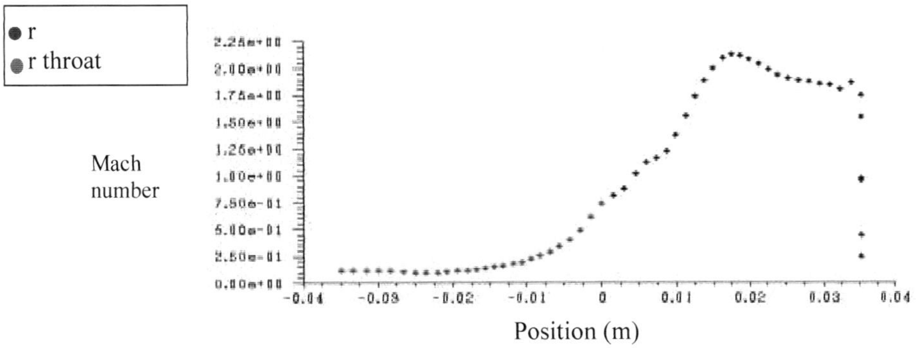


**Figure 22.** Mach No Vs Ramp wall plot of the 40% Truncated base bleed Aerospike Nozzle.

## 4. MANUFACTURING

### *4.1. FABRICATION MATERIAL:*

Stainless steel is an iron-containing alloy. It is a substance made up of two or more chemical elements used in a wide range of applications. It has excellent resistance to stain or rut due to its chromium content, usually from 12 to 20 percent of the alloy.
Stainless Steel 304 type material is used to fabricate the design because of its properties listed below.

*Raw Material:*

Stainless steels are made of some of the basic elements found in the earth: iron ore, chromium, silicon, nickel, carbon, nitrogen, and manganese. Properties of the final alloy are tailored by varying the amounts of these elements. Nitrogen, for instance, improves tensile properties like ductility. It also improves corrosion resistance, which makes it valuable for use in duplex stainless steels.

*SS304 Notes:*

Type 304 stainless steel is a T300 Series Stainless Steel austenitic. It has a minimum of 18% chromium and 8% nickel, combined with a maximum of 0.08% carbon.

It is defined as a Chromium-Nickel austenitic alloy.

These are some of its characteristics:

- Forming and welding properties
- Corrosion/ oxidation resistance thanks to the chromium content
- Deep drawing quality
- Excellent toughness, even down to cryogenic temperatures which are defined as very low temperatures.
- Low temperature properties responding well to hardening by cold working. - Ease of cleaning, ease of fabrication, beauty of appearance

*Applications:*

- Architectural panelling, railings & trim
- Chemical containers, including for transport
- Heat Exchangers
- Woven or welded screens for mining, quarrying & water filtration
- Dyeing industry
- In the marine environment, because of it slightly higher strength and wear resistance than type 316 it is also used for nuts, bolts, screws, and other fasteners.

IOP Conf. Series: Materials Science and Engineering **247** (2017) 012008 doi:10.1088/1757-899X/247/1/012008

| **Properties** | **Value** |
|---|---|
| Density | 8.03 g/cc |
| Tensile Strength, Ultimate | 621 MPa (=90100psi) |
| Tensile Strength, Yield | 290 MPa (=42100psi) |
| Elongation at Break | 55% |
| Modulus of Elasticity | 193 GPa |
| Modulus of Elasticity | 78 GPa |
| Melt temperature | 1371- 1399 °C |

*4.2. MANUFACTURING PROCESS:*

The whole design of the full length aerospike nozzle was drawn by using CATIA v5 software and the machine draft design chart was prepared by using AutoCAD 2009.

The complete Assembly Model of the aerospike nozzle was shown below,

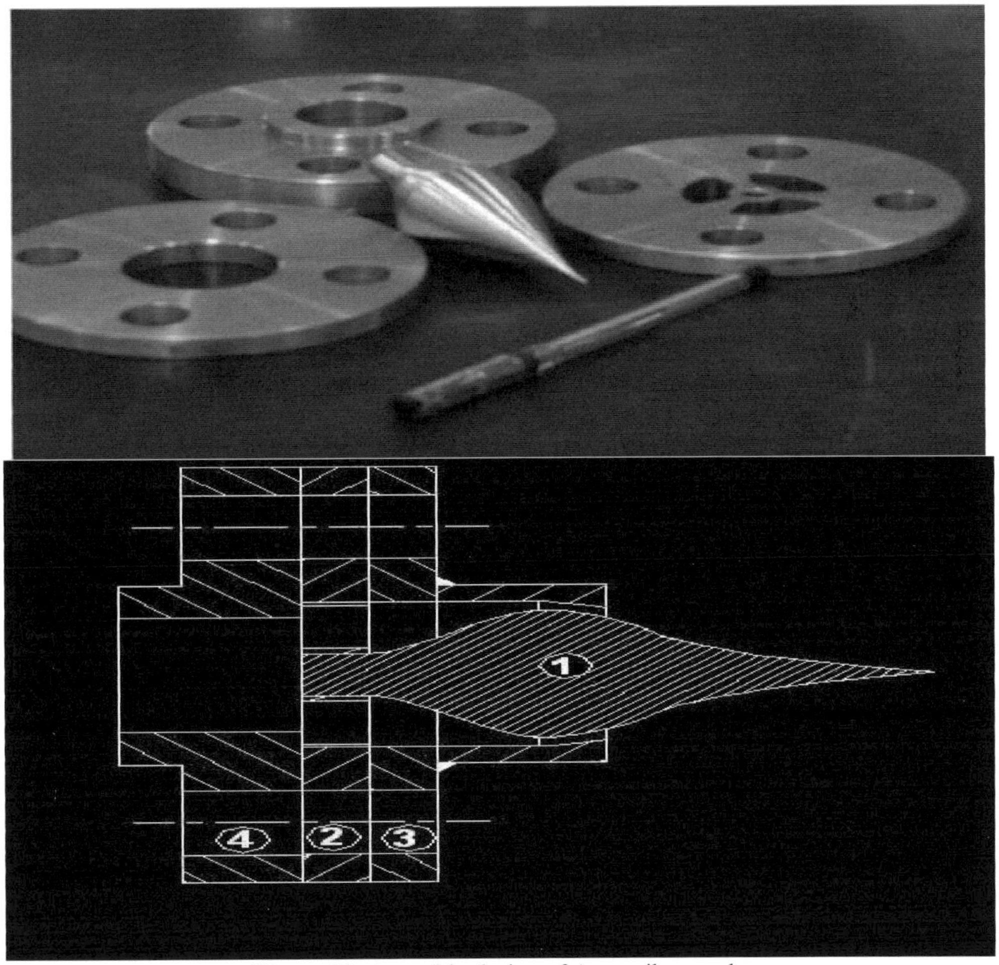


**Figure 23.** Assembly design of Aerospike nozzle.

1. *Spike:*

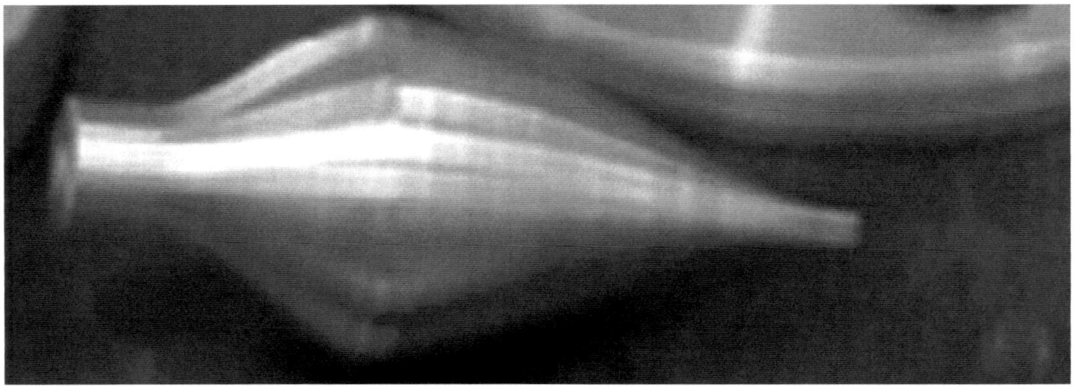

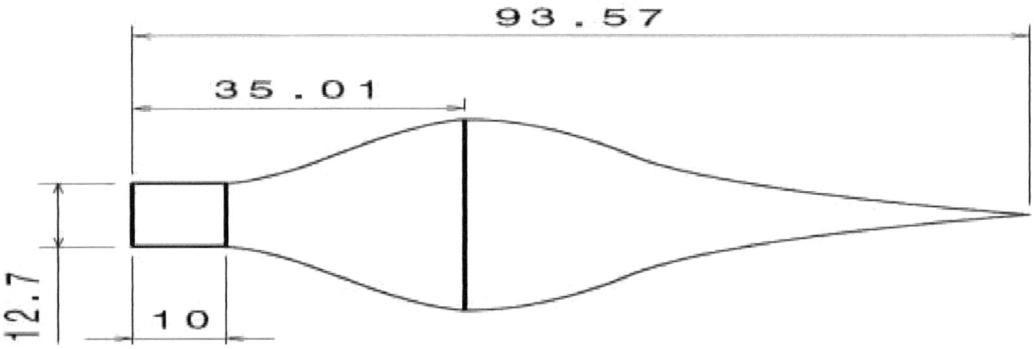


**Figure 24.** Design parameters of Spike

- The Spike part of the nozzle was created using CNC machine by installing the machine coding which includes coordinates of the spike.
- Additional length of 10mm of 12.75mm (1/2") dia was given for the model to fix with the plate. Clockwise male tapper threading was given on its surface.
- To create the spike 100X40mm piece of SS304 metal was taken.

2. *Plate*

- The important part of the nozzle is the plate; it is used to hold the spike.
- Clockwise female tap thread was created inside the centre hole.
- similar size 100 degree holes with the outer dia of 35mm. The holes allowed the air to enter the nozzle inlet.
- Hot wire cutting method is used to cut the metal for inlet holes.

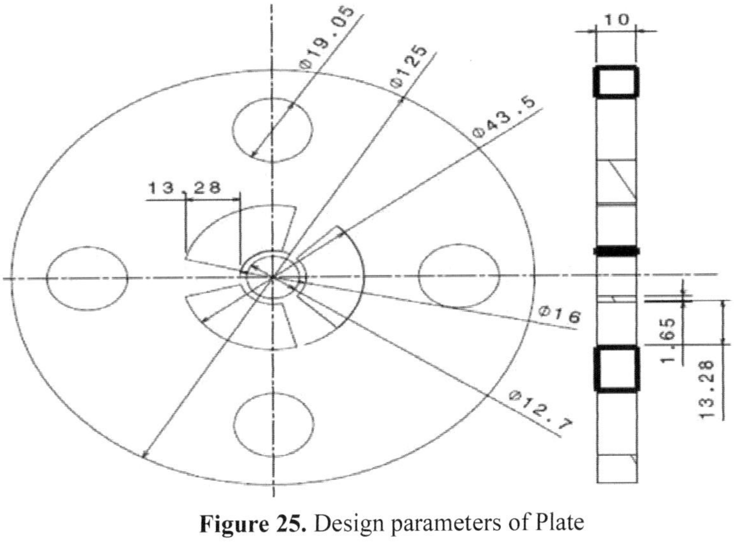


**Figure 25.** Design parameters of Plate

3. *Cowl and plate*
   - Initially the cowl was created using CNC machine in order to create the cowl lip arc.
   - The disc with the fitting holes created as shown in fig.
   - Finally, the cowl is forged with the plate.

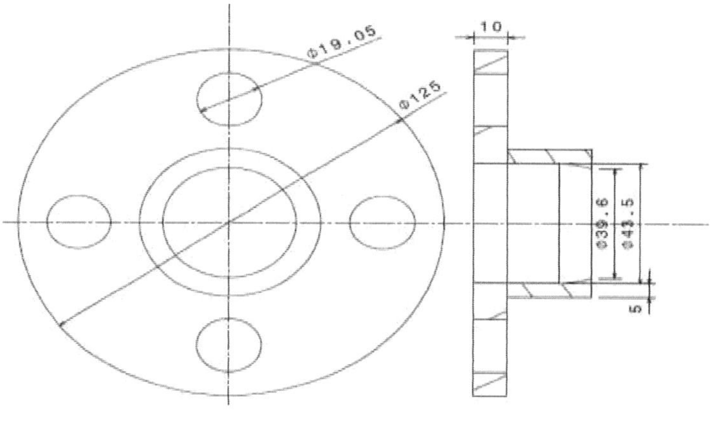


**Figure 26.** Design parameters of Cowl

4. *Flange:*

- 1inch 300# flange: The flange is readily available in all classes. Based on the size of the inlet valve of the wind tunnel and the nozzle operating pressure 300# (class) flange is used here.
- Since the flange has lesser Outer Dia (34.5mm) than the wind tunnel pressure inlet valve dia (38.4mm). The excess material will be removed from the flange.

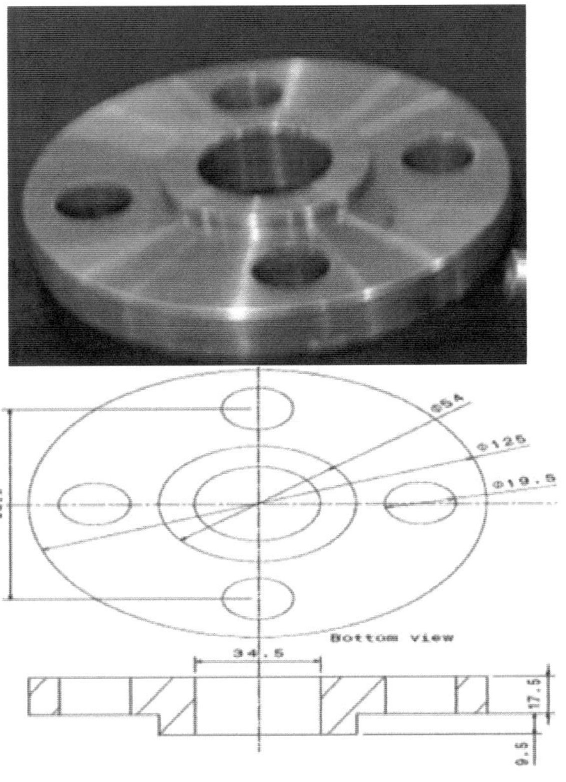


**Figure 27.** Design parameters of flange

## 5. TESTING OF PROTOTYPE

The fabricated prototype was tested in the pressure wind tunnel and the results were taken out for comparison.

### *5.1. Wind Tunnel Configuration:*

| Operating Parameters | Value |
|---|---|
| Air storage capacity | 8000 cubic feet |
| Max. Storage Pressure | 100 Psi |
| Air Compression Rate | 16 to 18lb/sec |
| Run time | 15 to 80 sec |
| Nozzle | 5 feet long |
| No. of adjustable jacks | 12 |
| Mach Number Range | Transonic .3 to 1.8 |

| | |
|---|---|
| | Supersonic 1.6 to 4.8 |
| Dynamic Pressure Range | 150 to 5000 PSF |

*5.2. Testing*

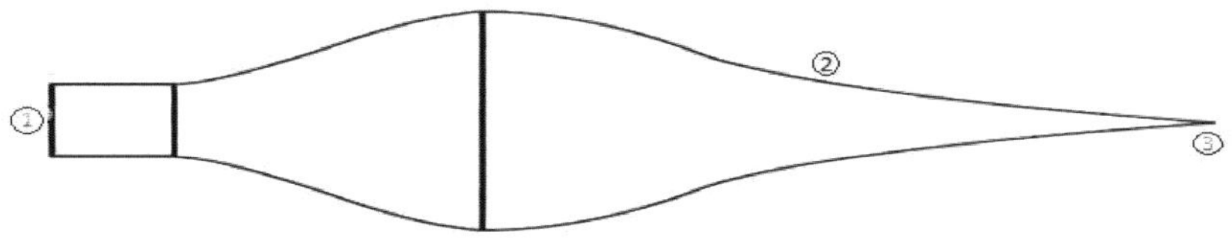

**Figure 28.** Pressure tap positioning.

Since it is a small model there are some difficulties on the insertion of the pressure taps. The removal of material inside the spike with the prescribed thickness became tough and the size of the taps available at the labs are 2mm so the number of taps were decreased into three.

1. At the beginning of the spike to measure the pressure at the combustion chamber (here from the pressure tunnel).
2. At the point of 40% of the ramp curve axis length (35.1402mm from throat).
3. Near the end of the spike (the point where the wake started in the computational analysis).

Due to some restrictions the images of fabricated model testing cannot be shown. But the results taken from the experiments were allowed to use and it is used to compare the results taken from the computational methods.

## 6. RESULTS AND DISCUSSION

Numerical modelling and analysis of internal and external flow of different aerospike nozzle shapes introduced in previous sections are described here. The objective of the analysis is to verify the prescribed exit flow properties of the nozzle design and compare the flow parameters between the nozzle designs.

*6.1. Mach Number Plots:*

Figure 29 shows typical Mach No plots of Experimental result of the full spike model and FLUENT generated graphs of 100% and 40% truncated and 40% base bleed aerospike nozzles designed for an exit Mach number of 2.601.

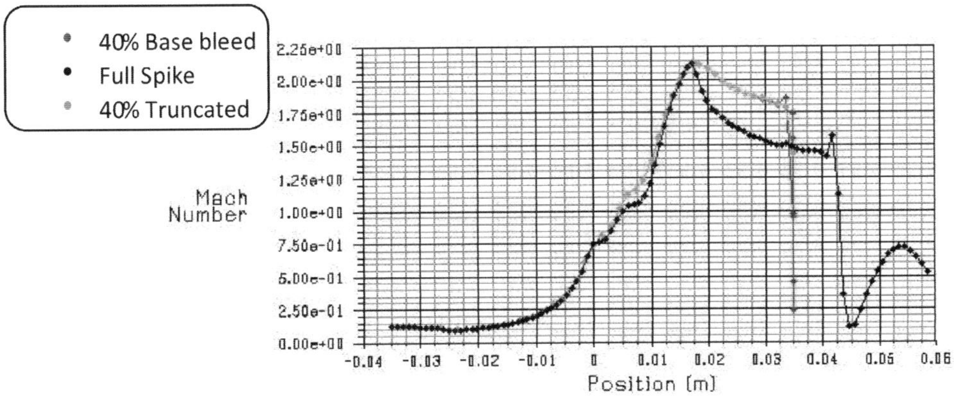


**Figure 29.** Mach No Vs Position of the ramp contour of the Full length & optimized nozzle models.

The maximum Mach number attained of full length aerospike nozzle is **3.82** and 40% truncated aerospike nozzle exit Mach number is **3.54** and the 40% truncated base bleed aerospike nozzle exit Mach number is **4.47**.

Although agreement between desired exit Mach number and simulated exit Mach number are below expectations, it is evident that the code is still relatively valid and that adjustment and refinement is needed to more accurately design an aerospike contour for a desired exit Mach number.

*6.2. Contours of pressure:*

*6.2.1. Dynamic Pressure contours:*

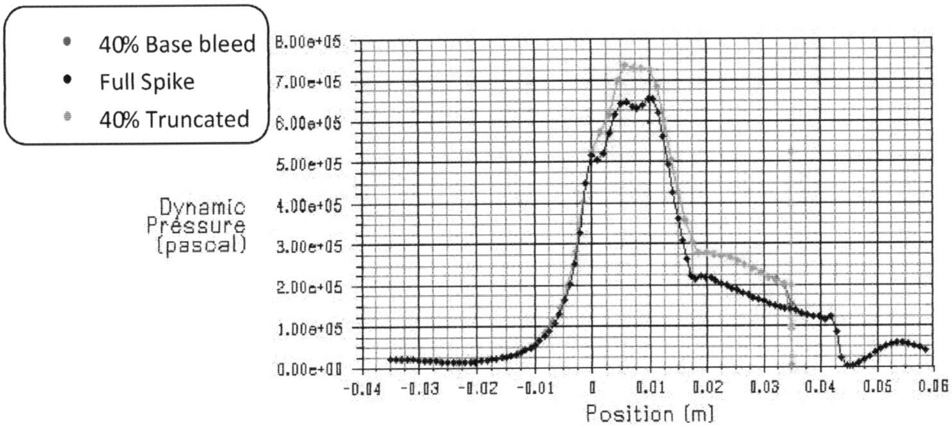


**Figure 30. Dynamic Pressure Vs Position of the ramp contour of the Full length, 40% truncated & 40% truncated aerospike nozzle with base bleed.**

Figure 30 shows the dynamic pressure variation throughout the nozzle. It seems that the dynamic pressure over the ramp up to truncated portion is having same value. The sudden decrement in dynamic pressure in 40% nozzle is due to the wake formation on the base and it was rectified by the 40% base bleed model.

## 7. Conclusion

The procedure to design an aerospike nozzle and design parameters that governs the aerospike nozzle design is discussed here. The designed value of the exit velocity (Mach No 2.5) is achieved with only 2.3% of error. The exit velocity of the air from FLUENT calculation is 2.44.

A comparison between the results of experimental and computational analysis of aerospike nozzle and also the performance of a full-length aerospike nozzle, a nozzle truncated at 60% of the full nozzle length& the same with the base bleed effect were done. For a single flow and boundary condition, the maximum Mach No attained at the end of the 40% truncated & Base bleed spike nozzle is 3.54 & 4.47 respectively.

i) The 60% truncated nozzle produces only a **10.8% lesser thrust** than the full-length nozzle. But it **reduces nearly 23% of the weight** of the full-length nozzle. The performance of the truncated nozzle will be more effective at higher altitudes. Since, the length of the nozzle is very less than the full-length nozzle, it produces considerable amount of performance. These findings are important when it comes to designing air and spacecraft since a nozzle with a lighter weight and equal performance is more attractive than a heavier nozzle.

ii) On the other side, the 40% truncated base bleed produces **14.35% more thrust** than the full-length nozzle and it also reduces around **29% of the weight** of the full-length nozzle.

But the only constraints of the base bleeding models are its additional mass flow rate. These nozzles need more mass flow rate for their base bleed flow (shown in table 4.2). We can use the base bleed models for high speed rocket engine applications. The diameter of the base bleed model can be modified based on the speed requirement. The diameter changes can be done only in ascending order. Experiments can be done for the optimization models and compared with the FLUENT results.

**References**

[1] Angelino G., "Approximation Method for Plug Nozzle Design", AIAA Journal, Vol. 2, No. 10, Oct.1964, pp. 1834-1835.

[2] Lee, C. C., "Computation of plug nozzle contours by the Rao's optimum thrust method", NASA CR-21914 R-61, 1963.

[3] Paul V. Tartabini, Roger A. Lepsch, Korte, J. J., and Kathryn E. Wurster, 2000, "A Multidisciplinary Performance Analysis of A Lifting-Body Single-Stageto-Orbit Vehicle", AIAA-2000-1045.Trans. Roy. Soc. London, vol. A247, pp. 529-551, April 1955.

[4] Verma, S. B., 2009, "Study of Restricted Shock Separation Phenomenon in a Thrust Optimized Parabolic Nozzle", J. Propulsion and Power.

[5] Shannon D. Eilers, Matthew D. Wilson, and Stephen A. Whitmore, 2010, "Analytical and Experimental Evaluation of Aerodynamic Thrust Vectoring on an Aerospike Nozzle", 46th AIAA/ASME/SAE/ASEE Joint Propulsion Conf. & Exhibit, Nashville, TN.

[6] Eilers S.D., Matthew D., Wilson and Whitmore S.A. (2010), 'Analytical and experimental evaluation of aerodynamic thrust vectoring on an aerospike nozzle',Joint Propulsion & International Energy Conversion Engineering Conferences, AIAA-jpc-10.

[7] Li Junwei., Liu Yu., LiaoYunfei and Wang Changhui (2010), 'Experimental and numerical study on two dimensional plug nozzle', 46th AIAA/ASME/SAE/ASEE Joint Propulsion Conference & Exhibit, AIAA paper 2010-6659, 2010.

[8] Takeo Tomita., Nobuhiko Kumada and Akira Ogawara (2010), 'A conceptual system design study for a linear aerospike engine applied to a future SSTO vehicle', 46th AIAA/ASME/SAE/ASEE Joint Propulsion Conference & Exhibit, AIAA paper 2010-7060,2010.

[9] Johnson G.R., Doyle Thompson H. and Hoffman J.D (1974), 'Design of maximum thrust plug nozzles with variable inlet geometry', Computers and Fluids, Vol 2, pp 173-190.

[10] Korte J.J. (2000), 'Parametric model of an aerospike rocket engine', AIAA paper 2000-1044, 2000.

[11] He Miaosheng, Qin Lizi, Liu Yu (2015), 'Numerical investigation of flow seperation behavior in an over-expanded annular conical aerospike nozzle', Chinese Journal of Aeronautics, (2015), 28(4): 983-1002.

[12] Johnson G.R., Doyle Thompson H. and Hoffman J.D (1974), 'Design of maximum thrust plug nozzles with variable inlet geometry', Computers and Fluids, Vol 2, pp 173-190.

[13] Korte J.J. (2000), 'Parametric model of an aerospike rocket engine', AIAA paper 2000-1044, 2000.

# A Survey on Friction Stir Welding Of Dissimilar Magnesium Alloys

Unnikrishnan M A[1]
[1]Department of Mechanical Engineering,
Noorul Islam University, Thuckalay, 629180
Tamil nadu ,India
Email id: unni037@yahoo.com

Edwin Raja Dhas.J[2]
[2]Department of Automobile Engineering
Noorul Islam University, Thuckalay, 629180
Tamil nadu ,India
Email id : edwinrajadhas@rediffmail.com.

***Abstract* -There is a consistent demand for superior materials in every industry. The areas on demand are automobile and aerospace sectors in major.. The most commonly used material in these fields is Aluminium.Though it possess all the properties up to some extent constant demand is pushing for alternate materials. Dissimilar alloys have been a relatively new approach towards these fields. . Friction stir welding dissimilar alloys is a big leap in Automobile sector. In this paper a detailed review of Friction stir welding of Dissimilar Magnesium alloys has been done. This work will serve as a reference to subsequent researchers.**

***Keywords:* *Friction stir welding, Magnesium alloy, Process parameter***

## I. INTRODUCTION

Friction stir welding (FSW) is a solid state welding process in which the rotary motion of a cylindrical tool along the welding line is the prime factor for the weld formation. The simultaneous rotation and translation of the tool induce a strong plastic deformation and promote complex mixing across the joint. The heat generated during the rotation of the tool will cause the materials to get joined without reaching melting point.[1]

## II. FRICTION STIR WELDING

Friction Stir Welding (FSW) was first invented by The Welding Institute (TWI), UK in 1991.This technique was developed aiming Aluminium alloys but later it had found profound application in welding of Mg alloys, Cu alloys etc.[2]FSW process consists of a rotating tool with a shoulder and probe arrangement. The heat is developed due to friction between the work piece surface and the tool [2]. The heat thus produced is used to soften the work piece before reaching its melting point. The heat generated during the process is about 80-90% of the melting temperature. With FSW traditional components current and voltage are not present as the heat input is purely mechanical replaced by force, friction etc. The quality of an FSW joint is always better than other fusion welding processes. In this process the FSW material consists of four distinct microstructural zones namely Nugget zone (NZ),Thermo mechanically affected zone(TMAZ),Heat affected zone (HAZ) and Base material (BM).The process parameters chosen during FSW process has great influence on these zones.[3]

## III. PROCESS PARAMETERS

The welding parameters are key players during Friction stir welding. Proper selection of welding parameters influences the final weld quality and resulting microstructure[4]. Some of the major process parameters are listed below.

### *A. Tool rotational speed and Travel speed*

This is one of the major process parameters during Friction stir welding and is primarily responsible for the generation of heat. The rotation may be clockwise or counterclockwise accordingly. The rotational motion of the tool generates frictional heat within the work pieces, and the material is brought to plasticized state and due application of pressure will cause the weld to form. The tool travel speed also plays a major role. It depends on several factors, such as alloy type, rotational speed, penetration depth, and joint type.[5]

### *B. Tool tilt and plunge depth*

A suitable tilt angle on the spindle provides an additional benefit for holding the stirred material and also for proper movement of plasticized material. The tilt angle is represented by (θ).The influence of plunge depth on formation of sound weld s are characterized by its influence on temperature distribution and on material mixing. The plunge depth of pin into the work pieces is important for producing sound welds with smooth tool shoulders [5]

*C. Tool characteristics*

Tool designs have influence on heat generation, plastic flow, the power required, and the uniformity of the welded joint. Tool profile which includes tool shoulder and pin is contributing for almost all factors influencing the weld formation. The variation in pin length and pin profiles has direct impact on material mixing, flow etc. The commonly used pin profiles are cylindrical, conical, threaded, square, octagonal etc.[5] The commonly used tool materials for FSW are Tool steel, carbide tool, High speed steel etc.

## IV. WELD QUALITY PARAMETERS AND MEASUREMENT

There are various quality parameters for a welded joint and these parameters are the judging factor for a good weld. Every welded joint must possess these quality parameters. The weld quality parameters are classified as

*A. Mechanical properties*

The mechanical properties of the weld are playing a crucial in a resultant weld Some of the common mechanical properties that determine the weld quality are Tensile strength , Strength/weight ratio, Elastic properties, Shear strength, hardness which Includes, micro hardness values, Ductility, Impact strength, percentage elongation,Wear,Corrosive behavior, fracture charecteristics,Fatigue etc.

*B. Microstructural properties*

. Every process parameter selected during welding has significant impact on final microstructure. All these can be found out by some common metallographic analysis techniques. In the case of dissimilar alloys welding one of the major limitation is the formation of intermetallic compounds. All these can be detected through proper microstructural and metallographic techniques. Some of them are Scanning Electron Microscopy(SEM),Surface topography, Dispersive X-ray analysis techniques which includes EDX & XRD,Optical microscopy etc.

*C. Thermal properties*

Apart from mechanical, microstructural properties on major influential parameter is temperature. The weld temperature varies during FSW owing to change in process parameters, tool pin profile etc. In case of dissimilar alloys the role of temperature is crucial as it determines the formation and thickness of intermetallic compounds. Some of the common thermal analysis methods are Thermo gravimetric analysis, Differential scanning calorimetry, Thermo mechanical analysis etc. All these analysis procedures are necessary for proper evaluation

## V. FRICTION STIR WELDING OF DISSIMILAR ALLOYS

Friction stir welding of dissimilar alloys are a growing area of concern. Though there is limitless scope yet there are some limitations also which has to be rectified. Some of the common hurdles during friction stir welding of dissimilar alloys are

*A. Heat generation and temperature distribution*

The discussion and analytical estimation of heat generation at tool work piece interface in similar welding is relatively easier because of the presence of only one type of material in contact with the tool. However, things become quite complicated when it comes to heat generation and temperature distribution in dissimilar FSW. Two different materials, in general, have different thermal properties, COF, and softening characteristics all these affect heat generation, temperature distribution in the work pieces, and material. The existence of an asymmetric thermal field would cause a sharp temperature gradient as a function of distance in lower thermal diffusivity material.. It might result in a lack of bonding between two materials which will lead to defects in the root region of the weld.

*B. Formation of intermetallic compounds*

The formation of intermetallic phases during dissimilar welding using conventional fusion welding is an issue because it can impair the joint integrity severely depending on the thickness. The formation of intermetallic compounds are greatly influence by welding parameters and the resultant temperature as well. Though the intermetallic formation is not completely avoidable in majority cases its effect , thickness etc can be reduced thus contributing to the final weld quality.

## VI. REVIEW ON FRICTION STIR WELDING OF DISSIMILAR MAGNESIUM ALLOYS

Shude Ji, Xiangchen Meng[6] researched on the formation of intermetallic compounds during Friction stir welding and its influence on Mechanical properties. The materials chosen were Aluminium alloy 6061-T6 and Magnesium alloy AZ31B plates of 3mm thickness. Tool rotational speed and welding speed were the major process parameters selected which was kept constant at 1200 rpm and 40mm/min.Ultrasonic was included along with the selected parameters. Detailed microstructural analysis were carried out using Scanning Electron Microscopy (SEM) and energy dispersive spectrometer (EDS).X-ray diffraction analysis shows the presence of $Al_3Mg_2$ and $Al_{12}Mg_{17}$ in the nugget zone. Fig 1 shows the microstructure details.

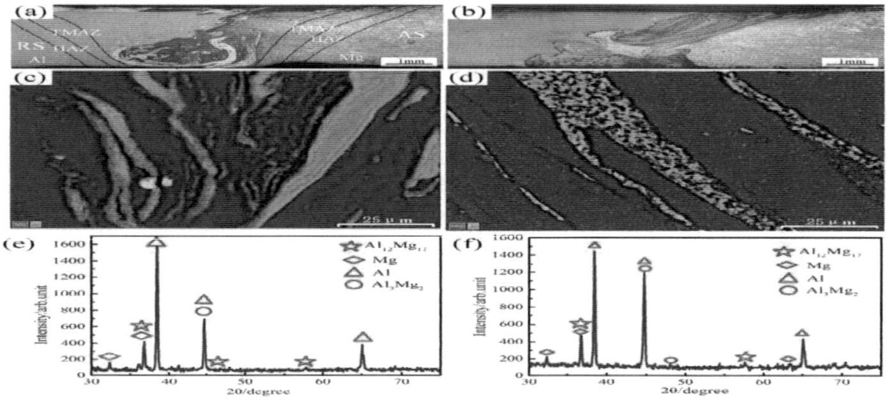


Fig 1. (a) and (b) shows the macrostructures with and without ultrasonic (c) and (d) shows the distribution map of Al/Mg elements with and without ultrasonic(e) and (f) shows the XRD patterns with and without ultrasonic

The average values of toughness for joints with and without ultrasonic are respectively $1.8Jm^{-3}$ and $0.3Jm^{-3}$, which shows the importance of ultrasonic on toughness.

V. Paradiso, F. Rubino[7] analyzed on the variation in mechanical as well as microstructural properties during friction stir welding of dissimilar alloys.The materials chosen for friction stir welding were ZE41A Mg alloy and AA2024-T3 Al alloy. Both the materials were of plate form with 4 mm thickness. Friction stir welding was carried out offsetting the tool of 1 mm towards the magnesium side. The tool was made up of high speed steel consists of a shoulder diameter of 20mm , a conical unthreaded pin of height 3.80 mm, major diameter 6.20 mm, and cone angle 30 deg. The process parameters selected were Tool rotational speed ranging from 1000 to 1400 rpm, feed rate ranging from 20 to 80 mm/min, tilt angle of 2° and shoulder plunge depth of 0.48 mm.

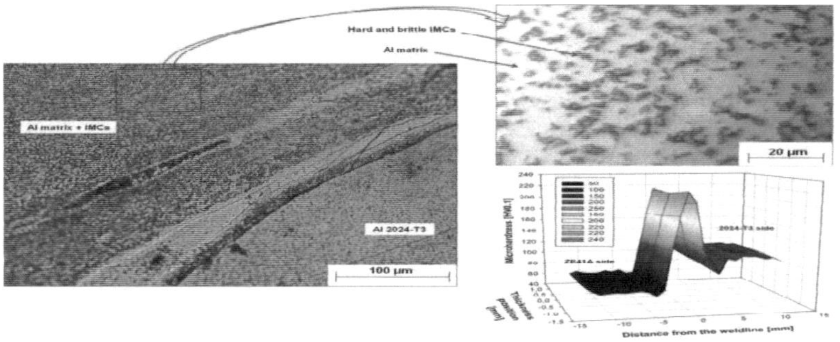


Fig. 2. Microstructural and micro hardness measurements of joint welded at 1200 rpm and 20 mm/min.

Yael Templeman and Guy Ben Hamu[8] studied about the microstructural evolution and corrosion resistance improvement during dissimilar alloys Friction stir welding. The materials chosen were AM50 and AZ31 Magnesium alloys. The process parameters were Tool rotational speed of 2500 rpm and welding speed of 0.1 mm/sec.Two pin diameters of 6mm and 2.7mm were selected accordingly. Scanning Electron Microscopy, Transmission Electron Microscopy and EDAX Energy Dispersive Spectroscopy (EDS) were employed for microstructural analysis. Phase identification was carried out X ray diffraction (XRD) diffractometer.Recrystallization was occurred during the process in the nugget zone.

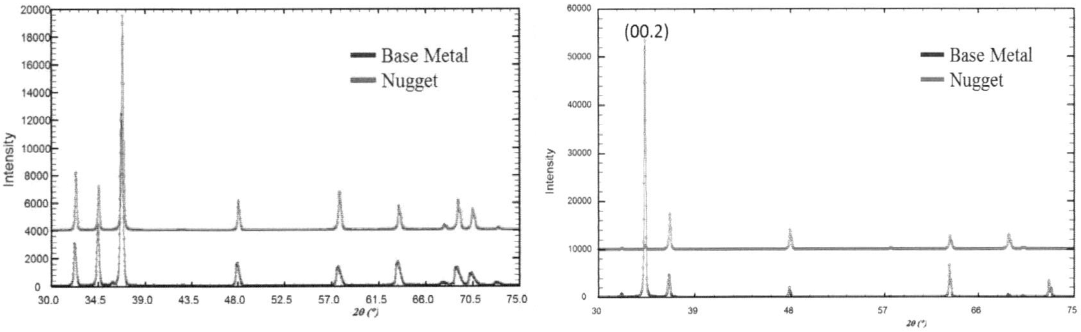


Fig 3(a) X-ray diffractograms of the base metal and nugget of the AM50 (b) that of AZ31.

The authors quote that in addition to the recrystallization, both Al-enrichment and β phase dissolved into the interior of the ά-Mg grains in the nugget area causing an increase in Al concentration and corrosion resistance improvement was noted in the nugget area.

Y. Zhao, L. Huang[9] investigated the effects of Tool travel speed on microstructure and mechanical properties of final weld. The materials chosen were Al 5754/AZ31 Mg alloy plates of 3mm thickness. The tool was made of H13 Tool steel with concave shoulder with diameter of 16 mm and a threaded pin with the length of 2.8 mm. and a constant tile angle of 3.5° was maintained Al alloy was placed on the AS and Mg alloy on the RS as per trials. Three kinds of defects had been detected during the experiment surface peeling, overflow of solidified microstructure and groove like defect are detected due to the variation in selected process parameters.

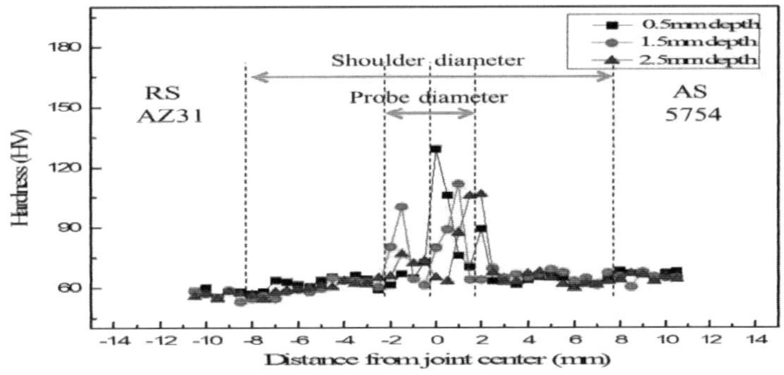


Fig 4 Microhardness profiles of microstructure from Mg to Al with different locations

SEM & EDS analysis has been carried out to found out the presence of intermetallic compounds, microscopic cracks which leads to a reduction in tensile strength. Also an uneven distribution of hardness values was detected from micro hardness studies. The author finds that defect-free weld joint could be obtained at a travel speed of 100 and 200 mm min−1, the maximum of tensile strength 76.6 MPa was obtained at 200 mm min−1 and it was 35% of AZ31 Mg alloy and A large amount of twins has been observed in the HAZ of the Mg side, and no HAZ was observed in theAl side. And no deformation grains occurred in the TMAZ of the Mg side of all the joints.

C.Luo, X.Li[10] investigated the microstructure and mechanical properties of dissimilar wrought Mg alloys AZ91D/ZG61plates of 6mm thicknesss.An adjustable pin tool made up of H13 Tool steel with a shoulder diameter of 15mm,pin length of 5.9mm with right hand thread was used. The process parameters were Tool rotational speed of 1000 rpm and welding speed ranging from 80mm/min to 240mm/min. Microstructural observation was done using SEM & EDS systems. Material flow analysis was carried out using electron backscatter diffraction (EBSD). The micro hardness of ZG61/AZ91D joints is higher than AZ91D/ ZG61 joints. Fine-grain strengthening and texture strengthening are two main observed strengthening mechanisms.

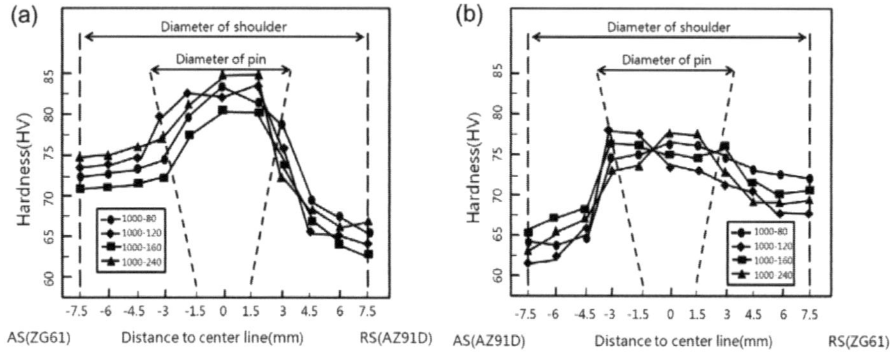


Fig 5. Microhardness profiles of the welds (a) ZG61/AZ91D and (b) AZ91D/ZG61

The author finds that severe deformation and metal flow occurred in the stir zone. During welding base metal is extruded into banded structure and vortex structure. The weld zone was harder than the base material due to smaller grain sizes in the stir zone

Heena K Sharma [11] in their investigation using circular butt joint geometry on AA6061/AZ31 Mg alloys used three different tool geometries along with tool rotational speed & tool travel speed as process parameters. The reasons quoted for not obtaining good quality weld were path of welding, pin geometry, higher values of process parameters. Circular path welding has been found to be inferior to linear path welding.

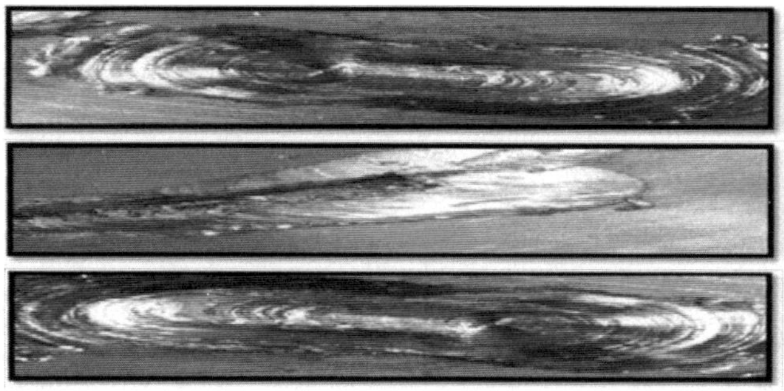

Fig 6.Circular butt joint welded specimen at a Tool rotational speed of 1200rpm and welding speed of 10mm/min

Five different pin profiles has been chosen namely cylindrical flute pin tool made of HSS tool steel, Threaded pin tool made of HCHCr ,Cylindrical groove pin tool made of H13 Tool steel, Cylindrical threaded pin tool and cylindrical threaded pin tool HCHCr. The tool rotational speed ranged from 800 rpm to 2000 rpm and tool travel speed ranged from 10 mm/min to 40 mm/min.The author finds that a tool rotational speed of 1200 rpm and welding speed of 10mm/min was the most effective set of parameters.

B. Ratna Sunil [12] had done investigations on dissimilar Mg alloys AZ31/AZ91 plates of 3mm thickness.The tool was made of H13 Tool steel with a shoulder diameter of 15mm,tapered pin with 3mm to 1mm diameterand 3mm length. The process parameters selected were tool rotational speeds of 1400rpm,1600rpm,1800 rpm and feed rate of 25mm/min,50mm/min,100mm/min

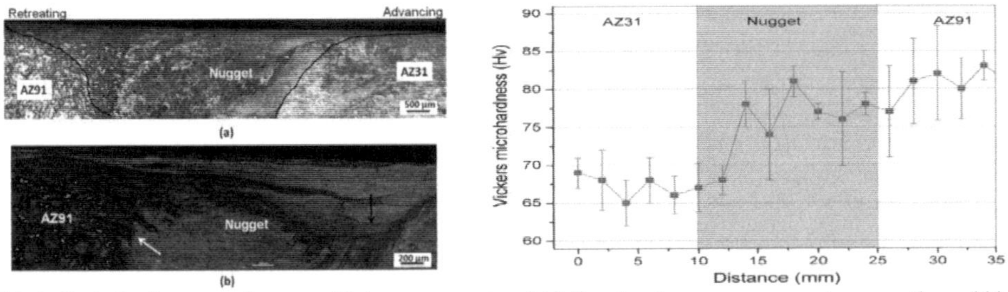


Fig 7 ( a ) Optical microscope images of joint cross section (b) Micro hardness measurements across the weld joint.

The amount of β (Mg17Al12) phase in AZ31 Mg alloy is lower compared with AZ91 Mg alloy.. The distribution of β (Mg17Al12) phase was limited to retreating side. Wire cut EDM was used for cutting the specimen after welding. The author states that increased hardness in the nugget zone can be attributed to the grain refinement and the presence of Mg17Al12 particles along with solid solution strengthening.

M. Tabasi [13] had concentrated on the Friction stir welding of dissimilar Al/Mg alloys. The alloys chosen were Al 7075 & AZ31 Mg having 5mm thickness.. silicon carbide nanoparticles were introduced into the weldment for the formation of metal matrix composites. Tool rotational speed of 450, 560, 710,900, and 1100 rpm and traverse speed of 11.2, 22.4, 35.5, and 45 mm/min were selected as process parameters. The tool selected for experimentation was of H13 Tool steel material with triangular threaded pin.

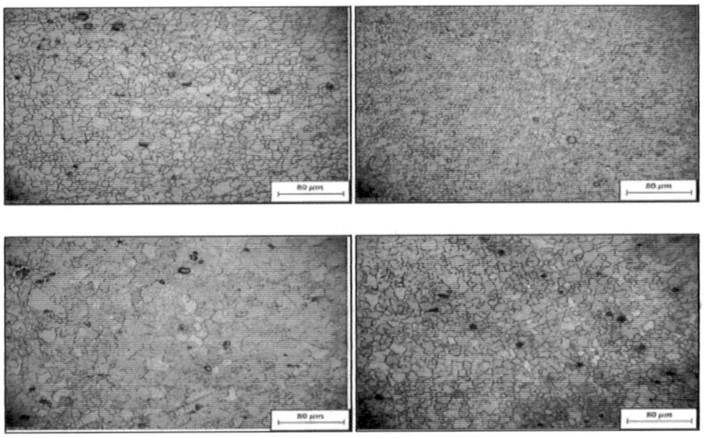

Fig 8. Microstructure of FS welded joints on the Mg side stir zone clockwise from (a) at TRS 560rpm and WS 22.4mm/min (b) TRS 710rpm and WS 22.4mm/min (c) TRS 560rpm and WS 35.5mm/min (d) TRS 710rpm and WS 35.5mm/min

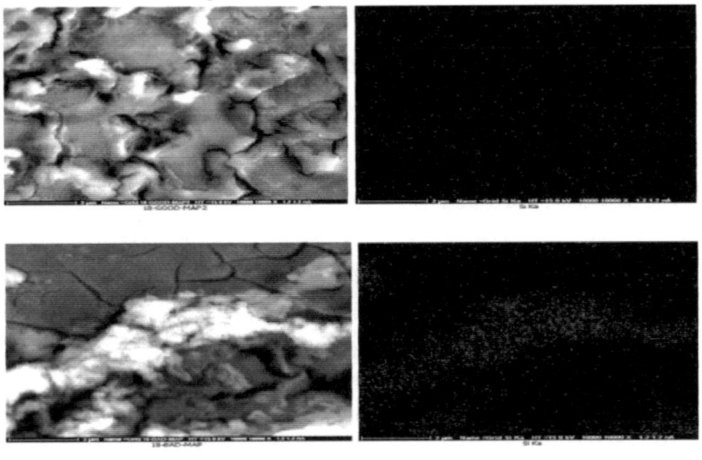

Fig 9. FESEM image and map of SiC distribution in the welded sample at (a) and (b) TRS 710 rpm and WS 22.4mm/min (c) and(d) TRS

Optical microscopy and Field Emission Scanning Electron Microscopy (FESEM) were used in order to analyze the microstructure of the stir zone and to study the particle distribution. Furthermore, energy-dispersive X-ray spectroscopy (EDAX) was employed for elemental analysis in stir zone of the specimens. Dynamic recrystallization occurred during the process which leads to nucleation & decrease in grain size.

Hui Shi [14] experimented on the effect of intermetallic compounds in the banded structure zone and its effects on mechanical and microstructural properties during friction stir welding with selected process parameters. The material chosen were Al 6061 & lab prepared Mg alloy. A conical pin profiled tool made up of H13 Tool steel was employed. The shoulder diameter was 12 mm, pin diameter of 4 mm and pin length of 2.5 mm, respectively. The tilt angle of the tool was set 3° forward and the plunge depth was 0.20 ± 0.05 mm during welding.

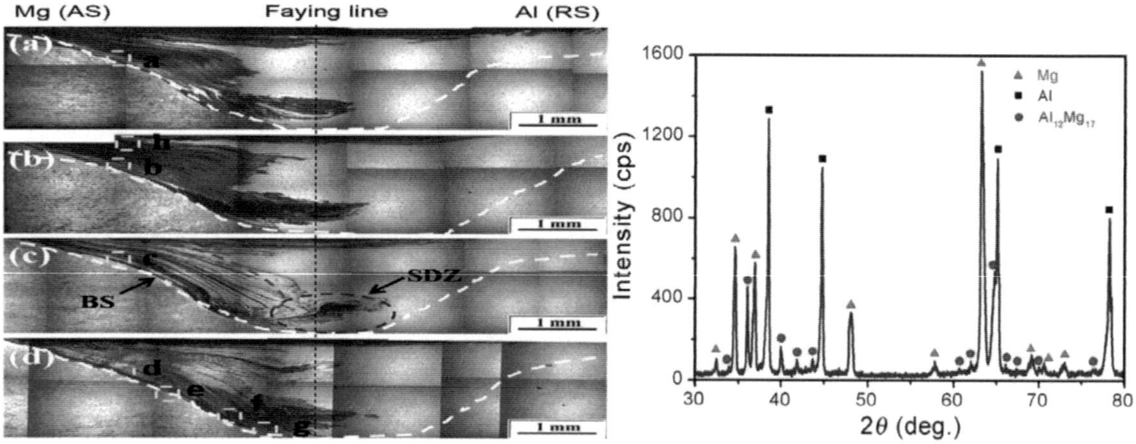


Fig 10 (a) Transverse cross sections of the welds at different TRS (i) 600 rpm (ii) 700 rpm (iii) 800 rpm and (iv) 900 rpm (b) XRD spectrum of polished fracture surface at 900 rpm.

Tool rotational speed was varied from 600-1000rpm & tool travel speed was fixed at 100mm/min. Energy Dispersive Spectroscopy (EDS) was utilized for microanalysis of chemical composition. IMCs in the BS were characterized by X-ray diffraction (XRD) method. According to the author BS microstructure should be tailored with the following characters for improving the strength of the BS zone: (i) less IMCparticles in the BS zone; (ii) bands in curved shape rather than straight arrangement; and (iii) bands being relatively short and discontinuous.

A. Dorbane [15] investigated on FS welding of dissimilar Aluminium AA6061-T6/AZ31 Mg alloys. The work concentrated on microstructural and mechanical aspects of weld joint. Optical microscopy and Scanning Electron Microscopy has been employed for microstructure analysis.

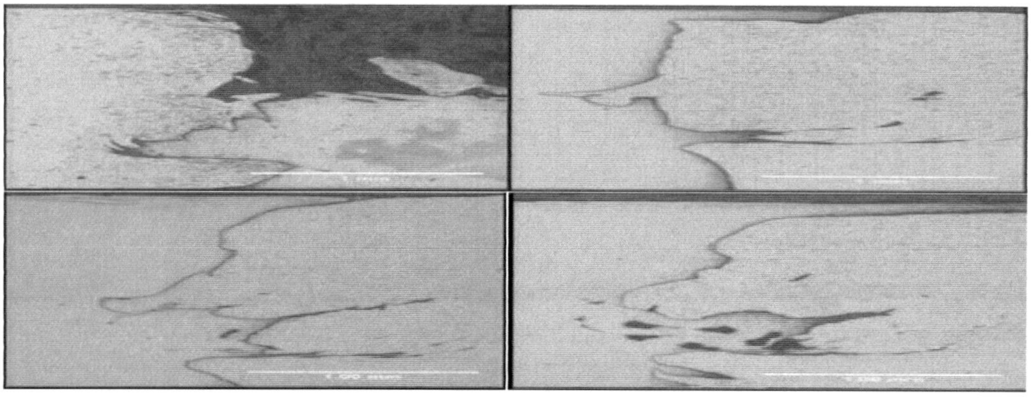

Fig 11.Optical microscope observation of the cross section of the welded joint (a) 1600 rpm and 250 mm/min translational speeds (b)to (d) Fixed welding speed of 500 mm/min and TRS of 1600, 1400, 1200rpm

Among the studied weld process parameters, 1400 rpm rotational speed and 500 mm/min translational speed resulted in the optimum joint quality, based on the analysis of the weld integrity and microstructure. The effect of mutual positioning on final weld quality has been found out. Grain refinement occurred and A thin layer of intermetallic compounds (IMC) were seen in the stir zone at the interface between the welded sheets.Microhardness tests and tensile tests had been done for validation. The process parameters chosen were Tool rotational speed (1200, 1400, 1600rpm) and a fixed tool travel speed of 500mm/min

M. Azizieh [16] investigated the effectof process parameters on final weld quality and also on intermetallic formation. The materials chosen were Aluminium AA1100/AZ31 Mg alloys. The tool selected was of H13 tool steel with cylindrical threaded pin. The researcher focused on offsetting the pin position towards Aluminium and Magnesium sides.

Fig 12. Onion ring pattern formation during welding

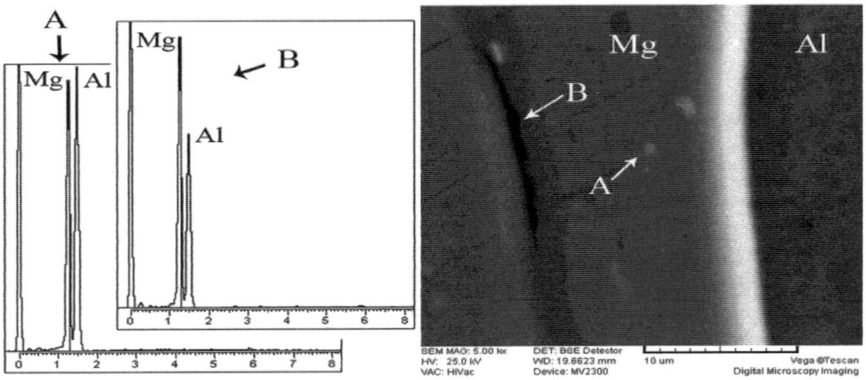


Fig 13. SEM macrograph and EDS results

The tool had a shoulder diameter of 18mm,threaded pin having a height of 2.7mm and diameter of 5mm.The threaded probes are of 1mm pitch.The pin position of the tool was set at five positions namely at the centerline,1.5mm off the centerline towards AZ31 ,2.5mm off the centerline towards AZ31,1.5mm off the centerline towards AA1100 and 2.5mm off the centerline towards AA1100.The FSW was conducted at several tool rotational speeds of 420, 480, 570, 660, 750, 800 and 1000 rpm and three travel speeds of 15, 20, and 30 mm /min. Optical and scanning electron microscopy (SEM) ,X-ray diffractometer (XRD) were employed for microstructural observation. For mechanical evaluations, transverse tensile and micro hardness tests were conducted..

Yong Zhao [17] researched about the influence of cooling conditions on joint properties and microstructure of Al/Mg dissimilar alloys. The materials chosen were Aluminium AA6013/AZ31 Mg alloys. Underwater friction stir welding and normal friction stir welding in air were conducted to study the influence on microstructure and mechanical properties of the weld joint.

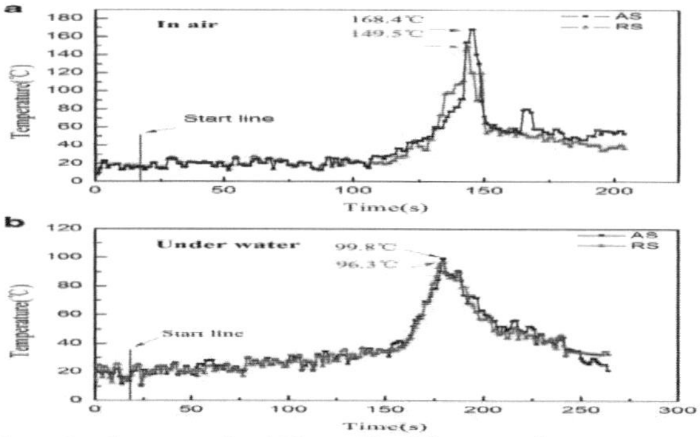


Fig 14. The thermal cycle curves of weld formed in different media. (a) In air (b) Underwater

Weld was made using H13 Tool steel with cylindrical pin profile, a concave 16- mm-diameter shoulder and a 5-mm-diameter pin with the length of 2.5 mm. The welding tool rotated counterclockwise and the tilt angle was 2.5° The Tool rotational speed and tool travel speed were fixed at 1200 rpm & 80mm/min.The thermal cycle temperature was measured by two thermocouples fixed on the work pieces. SEM was employed for fracture morphology analysis. The appearance of weld, microstructure and intermetallic details were better in under water FSW than in air.

J. Mohammadi [18] investigated on the microstructural and mechanical properties during FSW of AA6061/AZ31B Mg alloys. Rotation and travel speeds varied between 560-1400 r/min and 16-40 mm/min respectively and the tool was of H13 Tool steel with a threaded tapered pin.X-ray diffraction pattern (XRD), optical microscopy images (OM), electron probe microanalysis (EPMA) and scanning electron microscopy equipped with an energy-dispersive X-ray spectroscopy (SEM-EDS) were used to investigate the microstructures of the joints welded. Micro hardness tests and tensile tests had been performed after welding

H.M. Rao [19] researched on Friction stir spot welding Magnesium alloy AM60B and Aluminum alloy 6022-T4 under various welding conditions. The influences of tool rotation rate and shoulder plunge depth were examined. Welds were made at four different tool rotation rates of 1000, 1500, 2000 and 2500 rpm and various tool shoulder plunge depths from 0 mm to 0.9 mm.

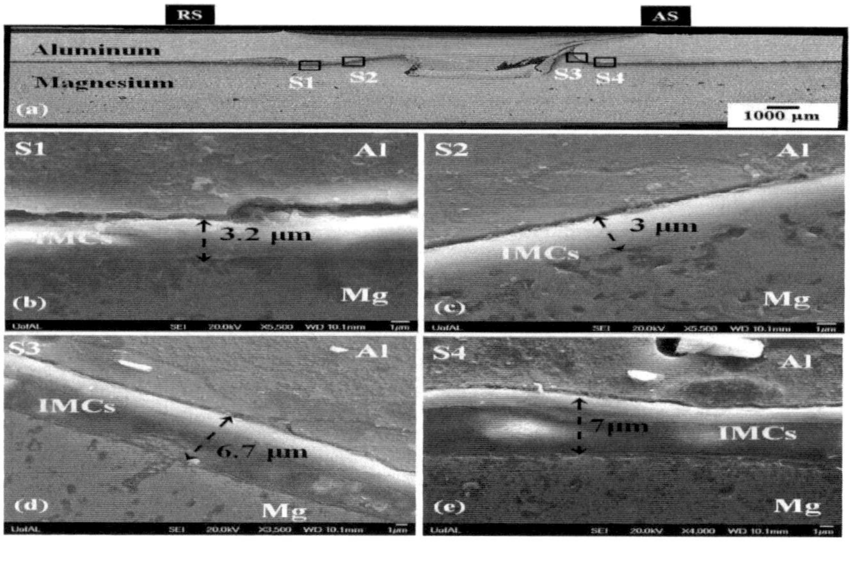


Fig 15 .SEM images of the IMCs on the faying surface in the Friction Stir Linear welding specimen

FSW tool used in this study was made of standard tool steel (H13) and constituted a concave tool shoulder and a triangular pin with threaded groove surface. A continuous layer of IMCs resulted in reduced weld strength while discontinuous IMCs resulted in improved weld strength.

Bang long Fu [20] investigated on the FSW of dissimilar Aluminium AA6061/AZ31B Mg alloys. The tool was made up of H13 Tool steel. The tool shoulder was on concave shape with 10mm diameter and frustum shaped right hand threaded pin. The macro and microstructure were examined using optical microscope, scanning electron microscope (SEM) equipped with EDX capability.

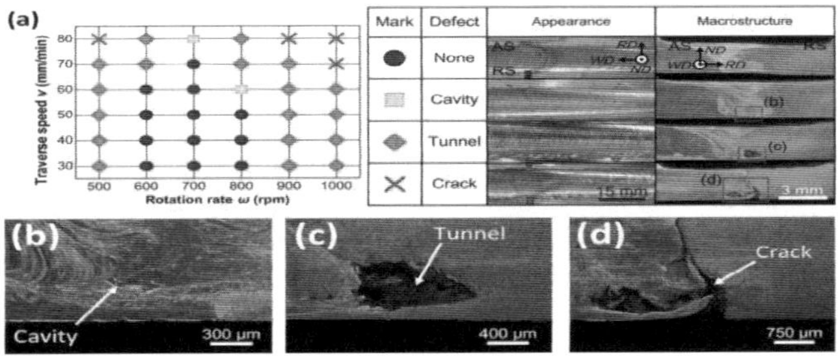


Fig 16.Weld appearance and microstructure of the joint with a tool offset amount of +3mm

A tool rotational speed from 600-800 rpm & tool traverse speed from 30–60 mm/min were chosen. Heat input in Al–Mg dissimilar FSW could be calculated from spindle and x-axis torque. The nugget zone at Mg side was divided into affected zone,banded zone and severe intercalated zone.The heat input shows an increase when the tool was offset towards aluminium side.The heat input during the process and proper intermixing contributed to better weld properties as dissimilar materials owe these two properties in major.

Kwang-Jin Lee [21] researched on the microstructural effects during FSW of dissimilar Al-Mg alloys. The materials chosen were AA6061-T6 and AZ31 alloys. The welding between AA 6061-T6 an AZ31 alloys were done with a tool offset towards AZ31 side . Electron back-scatter diffraction (EBSD) technique was applied to measure texture in the stir zone (SZ). Grain size distribution and misorientation angle distribution were also obtained.

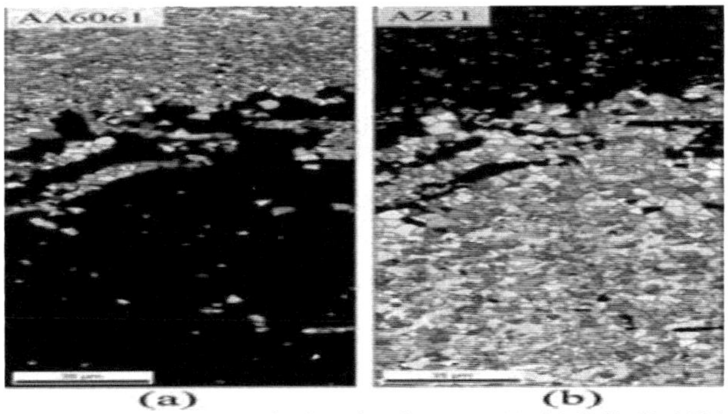


Fig 17. EBSD IPF maps for interface between (a) AA6061 (b) AZ31

Microstructure of the interface consisted of lamellar-like shear bands rich in either Mg or Al. The microstructure shows complex intercalated flow patterns. The tool with concave type shoulder and truncated cone type probe was made of general tool steel of SKD61. Tool rotation speed of 1200 rpm and the tool traveling speed of 10.16 cm/min was selected. The interface region was characterized by lamellar like shear bands.

Alireza Masaudian [22] dissimilar friction stir welding between AZ31-O Mg and 6061-T6 Al alloys was investigated. The rotation speeds varied from 600 to 1400 r/min, and the travel speed varied from 20 to 60 mm/min. An unthreaded cylindrical tool made from H13 tool steel. An unthreaded cylindrical tool made from H13 tool steel with a shoulder of 15 mm in diameter and concaved, a pin of 3 mm in diameter and2.9 mm in length, and a tilt angle of 2.5° was used for the welding operation.

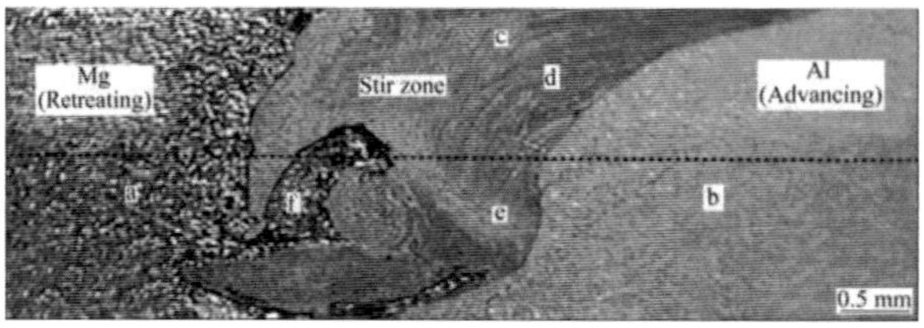


Fig 18.Transverse cross sectional micrograph of the weld

Al alloys was placed on the AS and Mg alloy on the RS. Elemental analysis of the weld and fracture surface fractography were performed using scanning electron microscopy (SEM)Intercalated microstructure was formed in some regions in the stir zone, and this complex flow pattern may be responsible for the uneven micro hardness distribution in the stir zone.

P. Venkateswaran [23] studied about the various factors that affect the weld quality during friction stir welding of Al/Mg dissimilar alloys. The materials chosen were AA 6063 aluminium and AZ31B Mg alloys. The welding was done using tool made up of H13 Tool steel with a fluted probe. Tool rotational speed from 900-2700rpm,Tool travel speed from 1.69-6.4mm/min and axial force from 14-30KN were selected

Fig 19.A continuous intermetallic layer along the weld interface

The microhardness distribution of the weld joints is mainly influenced by the local microstructural constituents such as the Al–Mg type intermetallic compounds, strengthening precipitates on the Al sides and grain size on the Mg sides of the weld joints.The elemental analysis of the fracture surfaces was done by energy dispersive spectroscopy (EDS) and microstructural analysis was carried out using Scanning electron microscopy (SEM). Transverse tensile strength of the weld joint is strongly influenced by the interface features.

Pooya Pourahmad [24] investigated about the materials flow and phase transformation in friction stir welding of Al 6013/Mg alloys. The tool material was of H13 Tool steel with a conical pin profile. The process parameters selected were tool rotational speed ranging from 800 to 2000 r/min, traverse speeds from 31 to 75 mm/min and tool tilt angle of 2° and 3° were applied

.

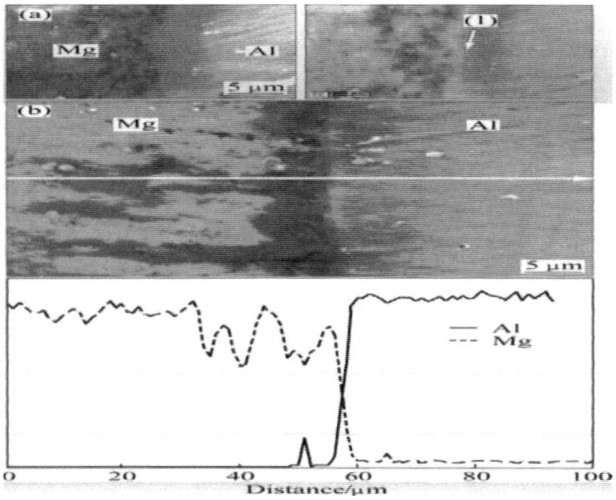


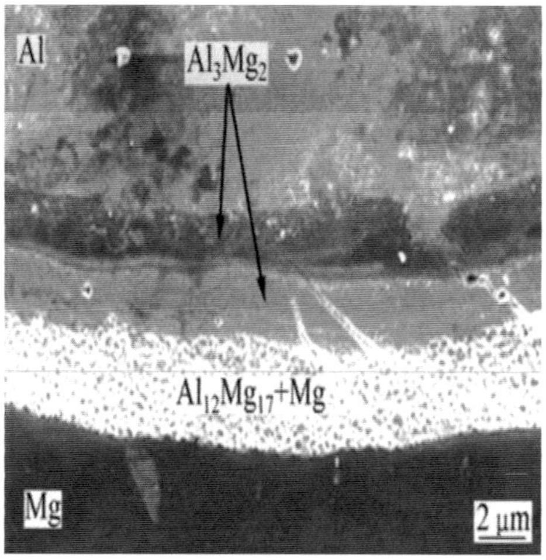


Fig 20. (a) SEM and EDS analysis of welded joints: (i) EDS mapping and quantitative analysis (ii) EDS line analysis from middle of Al/Mg interface

( b ) SEM image of Al/Mg interface in FSW joint after 4 h of heat treatment at 320 °C

The author quotes that defect free welds were obtained at a Tool rotational speed of 1600 rpm and welding speed of 35mm/min at a tilt angle of 3°.The microstructure of the dissimilar weld interface was analyzed by scanning electron microscopy (SEM) equipped with an energy-dispersive X-ray spectroscopy (EDS) analysis system. Micrographs of cross section of dissimilar welds revealed three regions. A zigzag interface, interlocking and islanding were observed in some regions . Regions showing extrusion of aluminium into magnesium and vice versa were observed.

W. Woo [25] investigated the influence of friction stir welding on changes in microstructures The process parameters tool travelling speed and rotating speeds were 4.7mm/sec, 1250 rpm (Al FSW) and 0.97 mm/sec, 600 rpm (Mg FSW) respectively. An H-13 steel tool was used for welding.

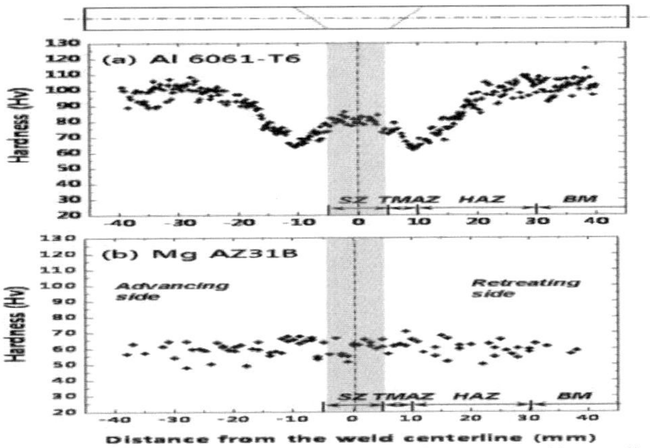

Fig21. Micro hardness measured on cross-section across weld centerline

The area of research by the author was concentrated on softening behavior and hardness related areas. Severe plastic deformation caused some texture variations which in turn resulted in softening of FSW Mg alloy. It results in a significant reduction in the yield strength and the residual stress in the SZ of the FSW Mg AZ31B alloy.

Yan Yong [26] investigated on the microstructural aspects during friction stir welding of dissimilar alloys. The materials chosen were Aluminium 5052 and AZ31 Magnesium alloys. During FSW process, 5052 aluminum alloy and AZ31 magnesium alloy were placed at the advancing side (AS) and the retreating side (RS) of the tool pin, respectively. A tool with a concaved shoulder of 15 mm in diameter and a cone-threaded pin of 6 mm in diameter and 5 mm in length was used. The tilt angle was 2° from the normal surface of plates

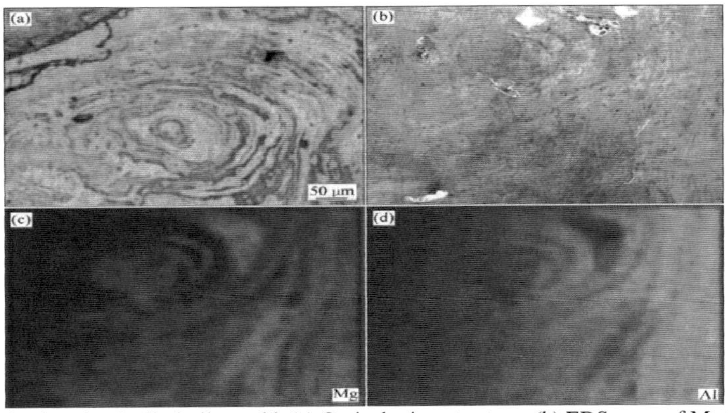

Fig. 22 Microstructures of onion ring in dissimilar weld: (a) Optical microstructure; (b) EDS maps of Mg (c) of Al (d) distribution in onion ring

The author finds that the microstructure of the base metal was replaced by equiaxed and fine grains in stir zone.The tool rotational speed was 600 rom and tool travel speed was 40mm/min. Micro hardness profiles presented uneven distributions and the maximum value of microhardness in the stir zone was twice higher than that of the base materials Microstructure of the weld was observed by optical microscope and scanning electron microscope equipped with an EDX system. The microstructural examination showed an increase in hardness at the stir zone than in the base material.

A. Kostka [27] researched about the microstructural analysis of friction stir welded Al-Mg dissimilar alloys.. The materials chosen were AZ31 magnesium alloy and AA 6040 aluminium alloy.

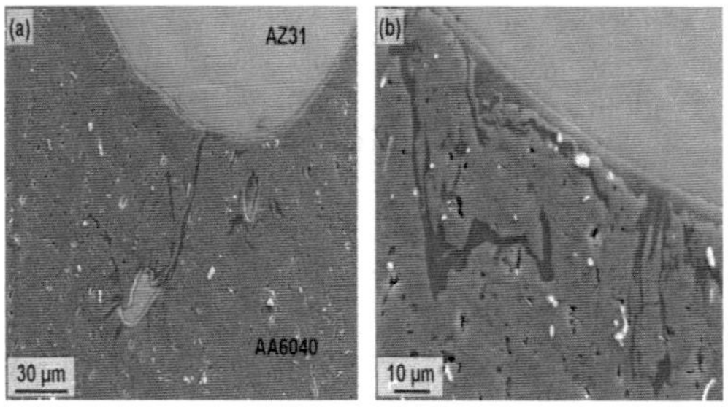


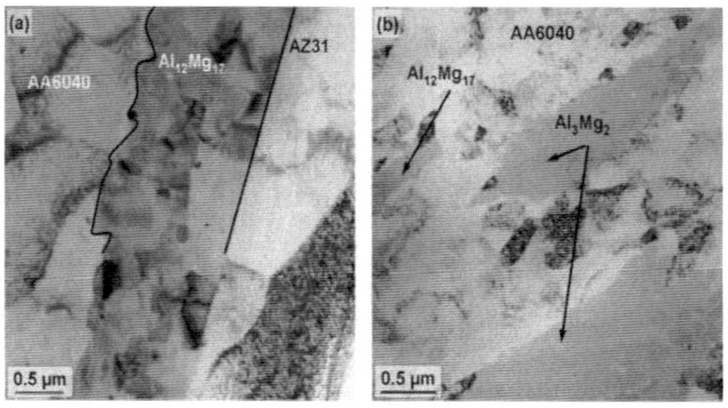


Fig 23. (a) Secondary electron micrograph of the AA6040 Al alloy and AZ31 Mg alloy interface (b) TEM micrograph of the interface region showing the intermetallic region

The process parameters were: rotation speed, 1400 rpm, travel speed, 3.75 mm /sec and axial force, of 3.5 kN.Scanning Electron Microscopy (SEM) examination of the interface microstructure revealed that intermetallic compounds occurred along the entire interface between the two alloys. The energy-dispersive X-ray SEM and TEM analysis of the joint were also done. The formation of the intermetallic compounds was studied in detail by TEM.

## VII. CONCLUSION

In this work a detailed review on Friction stir welding of dissimilar alloys has been considered. Though there are many the concentration of this work has been limited to dissimilar welding between aluminium and magnesium alloys and magnesium-magnesium alloys. Also the work is concentrated on butt joint Friction stir welding and related works. The influence of various process parameters on final weld quality, formation of intermetallic compounds and its influence on microstructure etc have been quoted. The works of various researchers have been highlighted throughout the paper.

## REFERENCES

[1] R.S.Mishra et.al ,Friction stir welding and processing,2005 Materials Science and Engineering R50 pages 1–78.
[2] A.Forcellese et.al , Effect of process parameters on vertical forces and temperatures developed during friction stir welding of magnesium alloys,2015 ,International journal of Advanced Manufacturing technology, Volume 85, Issue 1, pages 595–604.
[3] Prakash kumar sahu, Sukhomay pal , Multi response optimization of process parameters in friction stir welded AM20 Mg alloy by taguchi grey relational analysis,2015, Journal of Magnesium and alloys vol 3 pages 36-46.
[4] Unnikrishnan M A ,Edwin Raja Dhas.j, Friction Stir welding of Magnesium alloys-A review,2015, Advances in Materials Science and Engineering, MSEJ vol 2 no 4.
[5] Barath Raj singh, A handbook on Friction stir welding Research Gate,2012, DOI/10.13140/RG.2.1.5088.6244.
[6] Shude Ji .et.al , Dissimilar friction stir welding of 6061 aluminum alloy and AZ31magnesium alloy assisted with ultrasonic Materials Letters,2017, DOI/10.1016/j.matlet.2017.05.011.
[7] V. Paradiso.et.al , Magnesium and Aluminium alloys dissimilar joining by Friction Stir Welding,2017 , Procedia Engineering 183 pages239 – 244.
[8] Yael Templeman .et.al, Friction stir welded AM50 and AZ31 Mg alloys: Microstructural evolution and improved corrosion resistance, 2017, Materials Characterization, 10.1016/j.matchar.2017.02.018.
[9] Y. Zhao et.al, Effect of travel speed on the microstructure of Al- to-Mg FSW joints,2016, Material science and Technology.volume 32 issue 10.
[10] C. Luo et.al. Microstructure evolution and mechanical properties of friction stir welded dissimilar joints of Mg–Zn–Gd and Mg–Al Zn alloys,2016, Material Science and Engineering, A664(2016)103–113.
[11] Heena K Sharma et.al, Experimental Analysis of Friction Stir Welding of Dissimilar Alloys AA6061 and Mg AZ31 Using Circular Butt Joint Geometry,2016 Procedia Technology 23 ,pp 566 – 572.
[12] B. Ratna Sunil et.al. Joining of AZ31 and AZ91 Mg alloys by friction stir welding, 2015, Journal of Magnesium and Alloys 3 ,pp 330–334.
[13] M. Tabasi et.al. Dissimilar friction stir welding of 7075 aluminum alloy to AZ31 magnesium alloy using SiC nanoparticles,2015, The International Journal of Advanced Manufacturing Technology, Volume 86, Issue 1, pp 705–715.
[14] Hui Shi et.al, Intermetallic Compounds in the Banded Structure and their Effect on Mechanical Properties of Al/Mg Dissimilar Friction Stir Welding Joints,2016, Journal of Materials Science & Technology. Pages 1-8.
[15] Dorbane et.al, Mechanical, microstructural and fracture properties of dissimilar welds produced by friction stir welding of AZ31B and Al6061,2015,Materials science and Engineering ,DOI/10.1016/j.msea.2015.11.019.
[16] M.Azizieh et.al, Mechanical properties and microstructural evaluation of AA1100 to AZ31 dissimilar friction stir welds, 2015, Materials Chemistry and Physics ,volume 170,pages 251-260.
[17] Yong Zhao et.al .Influence of cooling conditions on joint properties and microstructures of aluminum and magnesium dissimilar alloys by friction stir welding,2015,International Journal of Advanced Manufacturing Technology,83: 673-679.

[18] J. Mohammadi et.al, Friction stir welding joint of dissimilar materials between AZ31B magnesium and 6061 aluminum alloys, 2015,Microstructure studies and mechanical characterizations,DOI/ 10.1016/j.matchar.2015.01.008.
[19] H.M. Rao et.al,Fatigue and Fracture of Friction Stir Linear Welded Dissimilar Aluminum-to-Magnesium Alloys, 2015, International Journal of fatigue, DOI/10.1016/j.ijfatigue.2015.09.033.
[20] Bang long Fu et.al, Friction stir welding process of dissimilar metals of 6061-T6aluminum alloy to AZ31B magnesium alloy,2015, Journal of Materials Processing Technology.218:38-47.
[21] Kwang-Jin Lee, Eui-Pyo Kwon ,Microstructure of stir zone in dissimilar friction stir welds of AA6061-T6 and AZ31 alloy sheets, 2014 Transactions of . Nonferrous Metals Society of .China 24: 2374−2379.
[22] Alireza Masoudian et.al.,Microstructure and mechanical properties of friction stir weld of dissimilar AZ31-O magnesium alloy to 6061-T6 aluminum alloy,2014, Transactions of . Nonferrous Metals Society of .China 2424: 1317−1322.
[23] P. Venkateswaran et.al . Factors affecting the properties of Friction Stir Welds between aluminum and magnesium alloys,2012, Materials Science and Engineering A.545:26-37.
[24] Pooya Pourahmed et.al ,Materials flow and phase transformation in friction stir welding of Al 6013/Mg,2013, Transactions of . Nonferrous Metals Society of .China 2423:1253−1261.
[25] W. Woo et.al ,Softening behavior of friction stir welded Al 6061-T6 and Mg AZ31B alloys,2011, Science and Technology of Welding and Joining 2011 vol 16 no 3.
[26] Yan Yong et.al, Dissimilar friction stir welding between 5052 aluminum alloy and AZ31 magnesium alloy,2010,Transaction of nonferrous metals society of china 20: s619-s623.
[27] A. Kostka et.al, Microstructure of friction stir welding of aluminium alloy to magnesium alloy,2009, Scripta Materialia 60 : 953− 956.
[28] Rajiv Mishra.et.al,A Framework for Friction Stir Welding of Dissimilar Alloys and Materials,2015,ISBN:9780128026212.
[29] Wei Guo et.al, Microstructure and mechanical properties of dissimilar inertia friction welding of 7A04 aluminum alloy to AZ31 magnesium alloy,2017, Journal of alloys and compounds, volume 695, pages 3267-3277.
[30] JFC Moraes et.al, Effect of overlap orientation on fatigue behavior in friction stir linear welds of magnesium alloy sheets,2017,International Journal of Fatigue volume 100 ,pages 1-11.
[31] Richard Johnson et.al,Friction Stir Welding of Magnesium Alloys,2016, Essential readings in Magnesium technology, part 8,pages 487-492.
[32] S.D.Ji et.al, Investigation of Ultrasonic Assisted Friction Stir Spot Welding of Magnesium Alloy to Aluminum Alloy,2016,volume 48,issue1,pages 2-7.
[33] B.Mansoor et.al., Friction Stir Welding of AZ31B Magnesium Alloy with 6061-T6 Aluminum Alloy: Influence of Processing Parameters on Microstructure and Mechanical Properties,2016,Friction stir welding and processing, pages 259-266.
[34] Todd.R.Curtis et.al, Joining Al 6061 to ZE41A Mg Alloy by Friction Stir Welding Using a Cold Spray Transition Joint,2016,Friction stir welding and processing ix,part vi pages 221-236.
[35] Zhida Liang et.al,Continuous drive friction welding of 5A33 Al alloy to AZ31B Mg alloy, 2017, Journal of Manufacturing processes, volume 25,pages 153-162.
[36] Uceu Suhuddin,et.al, Microstructure in Dissimilar Friction Spot Weld of Al to Mg Alloys Observed by Stop-Action Technique,2016,Friction stir welding and processing vii,pages 263-270.
[37] Zeina G. El Chlouk.et.al, The Effect of Tool Geometry on Material Mixing during Friction Stir Welding of Magnesium AZ31B Welds, 2015, proceedings of MEMA'15, pages 235-242.
[38] Zhe Liu .et.al, Textural variation in triple junction region of friction stir welded Mg alloys and its influence on twinning and fracture,2016,Material science and Engineering, volume 658,pages 185-191.
[39] A.Macwan.et.al, Similar and Dissimilar Ultrasonic Spot Welding of a Rare-Earth Containing ZEK100 Magnesium Alloy,2016,Magnesium Technology ,pages 109-113.
[40] S.M.Manladan.et.al. A review on resistance spot welding of magnesium alloys,2016, The International Journal of Manufacturing Technology, Volume 86, Issue 5, pages 1805–1825.

IOP Conf. Series: Materials Science and Engineering **247** (2017) 012010 doi:10.1088/1757-899X/247/1/012010

# *Mechanical Property Evaluation of Palm/Glass Sandwiched Fiber Reinforced Polymer Composite in Comparison with few natural composites*

Edwin Raja Dhas, J and Pradeep, P.
Department of Automobile Engineering,
Noorul Islam University, Kumaracoil,
Tamil Nadu, India.
edwinrajadhas@rediffmail.com, pspradeep2006@rediffmail.com,

*Abstract*— **Natural fibers available plenty can be used as reinforcements in development of eco friendly polymer composites. The less utilized palm leaf stalk fibers sandwiched with artificial glass fibers was researched in this work to have a better reinforcement in preparing a green composite. The commercially available polyester resin blend with coconut shell filler in nano form was used as matrix to sandwich these composites. Naturally available Fibers of palm leaf stalk, coconut leaf stalk, raffia and oil palm were extracted and treated with potassium permanganate solution which enhances the properties. For experimentation four different plates were fabricated using these fibers adopting hand lay-up method. These sandwiched composite plates are further machined to obtain ASTM standards Specimens which are mechanically tested as per standards. Experimental results reveal that the alkali treated palm leaf stalk fiber based polymer composite shows appreciable results than the others. Hence the developed composite can be recommended for fabrication of automobile parts.**

***Keywords***— *composites; palm-stalk; nano-filler; fiber-treatment; chopped-fibers;*

## I. INTRODUCTION

Natural fibers like palm, sisal, bamboo, kenaf, hemp, jute and coir are few bio-degradable natural fibers that are widely used in many automotive applications [1-3]. It is difficult to compromise the properties of natural fibers with synthetic fibers since their properties are closer to metals. Developing a green polymer composite using natural fibers are recent investigations among researchers as these polymer composites possess novel qualities like high tensile strength with less weight, low cost, easily available and holding good fiber matrix adhesion. Natural composites are presently alternatives for synthetic fiber composites [4-5].

Some naturally available fibers presently used as reinforcements in different thermosetting and thermoplastic resins were henequen, sisal, coconut fiber, jute, bamboo, date palm, wood, basalt and banana [6-13]. Many researchers manufactured composites with and without chemical treatment [14-15]. Fillers are added to natural fibers in preparation of composites for high strength [16-17]. Increase in percentage of coconut shell filler content usually increases the tensile strength.

Only inadequate research was done using palm fibers as reinforcement in natural composites. Palm fibers residues plus few percent of coupling agent enhanced the composite characters [18]. NaOH treated palm fruit fiber were analyzed using FEA for tensile prope les [19]. The developed Hybrid Oil Palm fibers had good mechanical characters and less water absorption properties [20]. The Sugar Palm Fiber reinforcement showed high tensile and impact strength [21]. Oil palm empty fruit bunch natural fibers with Copper nano-particles revealed biodegradability [22]. Brake pads for automotives are produced with palm kernel fibers [23]. Composite using long palm palm petiole fibers in polyester composites gave good mechanical property [24]. The graphite addition on date palm fibers in epoxy composites gave good mechanical and wear characteristics [25]. The alkali treatment improved the strength of hybrid kenaf/ Oil Palm Fruit bunch fibers. Hybrid jute with oil palm fibers reinforced using hand lay-up method finds application in automotive industry [26].

This paper addresses the development of palm leaf stalk fiber sandwiched with glass fiber filled with coconut shell nano filler reinforced polymer composites. Also the results from the treated palm leaf stalk fiber sandwiched with glass fiber filled with coconut shell nano filler reinforced polymer composites are compared with few natural fiber based composites and presented.

## II. EXPERIMENTATION

The materials used to develop the sandwiched composites are naturally available fibers extracted from the palm tree in its leaf stalk part as shown in fig1. To enhance the properties of these fibers 8% $KMnO_4$ solution was used. Commercially available E-Glass fiber mat was used to sandwich these composite. Coconut shell nano filler was added to fill in the voids formed during polymerization of the composites. To enhance the properties of these fibers 8% $KMnO_4$ solution was used. Commercially available E-Glass fiber mat was used to sandwich these composite. Coconut shell nano filler was added to fill in the voids formed during polymerization of the composites. Polyester resin with cobalt napthanate (hardener), methyl ethyl ketone peroxide (catalyst) forms the base or matrix material in the composites. Wax polish as mould releasing agent was used for mould surface coating.

Fig. 1. Palm leaf stalk fibers extracted from palm tree

### *2.1 Extraction of fiber and processing*

Palm leaf stalk taken from palm tree was soaked in water for two weeks and washed thoroughly to remove the impurities sticking on fibers. The fibers were then thoroughly washed with distilled water to remove further impurities followed by sun drying for about one week. The dried fibers are treated with 8% KMnO4 solution for about 3 hrs. This enhances the fibers properties suitable for adhesion. Finally the fibers are rolled as mat for reinforcement.

### *2.2 Preparation of specimen*

The Table 1 shows the materials used in various ratio for composite fabrication based on weight percentage. The composite plates A to D are fabricated in this combination. Composite plate A was fabricated with the treated **palm leaf stalk** fiber mat/sandwiched with E-glass fiber mat/ filled with coconut shell nano filler reinforced into the polyester resin that act as matrix. Composite plate B was fabricated with the treated **coconut leaf stalk** fiber mat/sandwiched with E-glass fiber mat/ filled with coconut shell nano filler reinforced into the matrix. Composite plate C was fabricated with the treated **raffia** fiber mat/sandwiched with E-glass fiber mat/ filled with coconut shell nano filler reinforced into the matrix. Composite plate D was fabricated with the treated **oil palm** fiber mat/sandwiched with E-glass fiber mat/ filled with coconut shell nano filler reinforced into the matrix.

Table 1 MATERIAL COMBINATION IN WEIGHT % FOR FABRICATION

| Specimen | Combinations | Weight % |
|---|---|---|
| A | Polyester resin + | 70 |
| | 8% KmnO4 treated **palm leaf stalk** fiber mat + E-glass fiber mat + Coconut shell nano filler | 30 |
| B | Polyester resin + | 70 |
| | 8% KmnO4 treated **coconut leaf stalk** fiber mat + E-glass fiber mat + Coconut shell nano filler | 30 |
| C | Polyester resin + | 70 |
| | 8% KmnO4 treated **raffia** fiber + E-glass fiber mat + Coconut shell nano filler | 30 |
| D | Polyester resin + | 70 |
| | 8% KmnO4 treated **oil palm** fiber + E-glass fiber mat + Coconut shell nano filler | 30 |

*2.3 Fabrication of Composites for testing:*

A mould box of required dimension is used to cast the composites. Hand lay-up method was adopted for fabricating all composite plates. Required amount of polyester resin was mixed with 2% cobalt napthanate and methyl ethyl ketone peroxide taken in a vessel, 5% coconut shell nano filler was also added and thoroughly mixed. The glass fiber mat was placed as first layer over the molding board and the required amount of polyester resin/coconut shell nano filler mix was poured over it. The trapped air bubbles were removed by gently rolling a hand roller over the fiber mat surface. The successive layers are made with natural fiber mats and the final layer was completed using glass fiber mat. The composites are cured in normal temperature and later removed from the mould box after curing. All the composite plates are sized as per ASTM D standards to performing various tests in order to determine the hardness values, tensile strength, flexural strength and impact strength of all composites.

## III. MECHANICAL CHARACTERISATION

*3.1 Hardness Value*

Rockwell hardness tester was used to compute the hardness values on all the specimens prepared as per ASTM D 785 as shown in the fig 2. Load of 60 Kg applied for 15 seconds on the specimen surface measures the hardness.

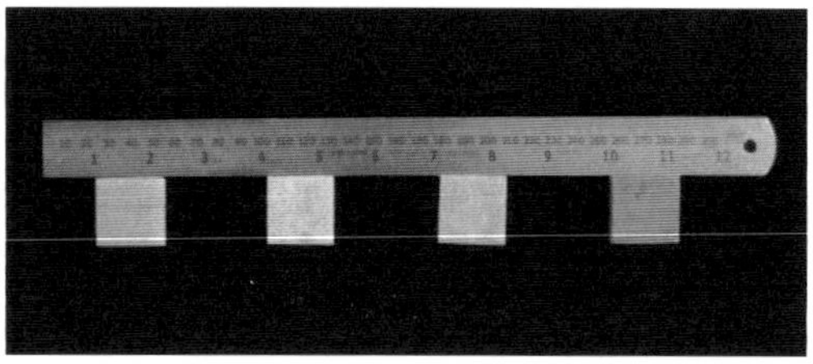

Fig. 2. Specimen for hardness test

*3.2. Tensile properties*

Tensile properties of 8% $KMnO_4$ treated natural/ glass fiber mat sandwiched/ coconut shell nano powder reinforced polymer composites are found by testing the samples of each material on a computerized servo controlled UTM machine with specimen standard ASTM D 638 as shown in fig 3 .The specimens are mounted in UTM and pulled apart for measuring strength and elongation.

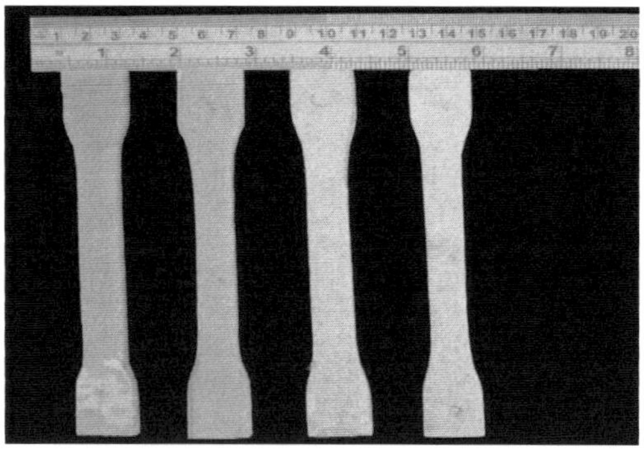

Fig. 3. Specimen for tensile test

*3.3 Flexural Properties*

The flexural properties of all samples are measured by conducting Flexural test on computerized UTM using special attachment with specimen standard ASTM D 790 as shown in fig 4 .The speed of test was set as 2mm/min at normal temperature.

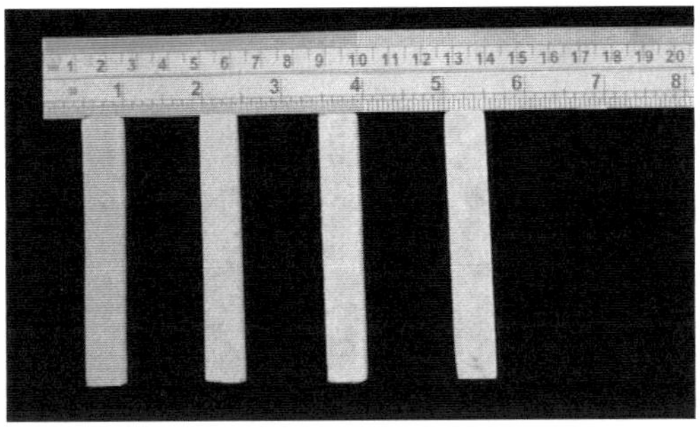

Fig. 4. Specimen for flexural test

### *3.4 Impact Strength*

The impact strength of all composites were measured in an charpy impact testing machine. The specimen standards adopted was ASTM D 6110 as shown in the fig 5. The fracture values are the ratio between energy absorbed to sectional area of the specimen.

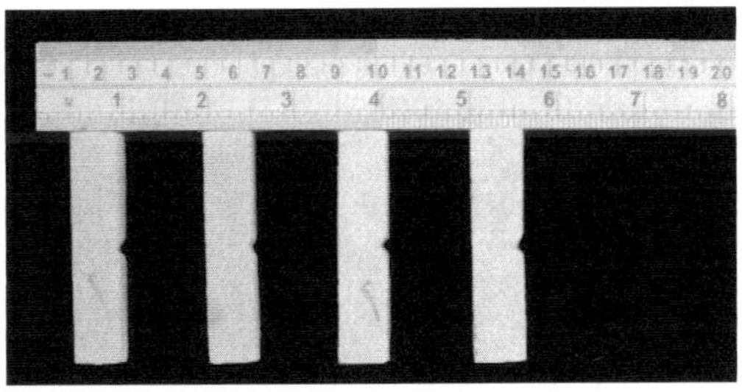

Fig. 5. Specimen for impact test

## IV. RESULTS AND DISCUSSION

The composite strength depends on the surface of the fibers and its interface with the matrix. Alkali treatment and addition of nano coconut shell filler material filled the gaps and decreased the voids formed during polymerization providing good fiber matrix adhesion. These characters are detailed as follows.

### *4.1 Specimen Vs Hardness*

The variation of hardness for all the tested specimens was shown in Fig 6. The specimen A recorded maximum value of hardness than the rest.

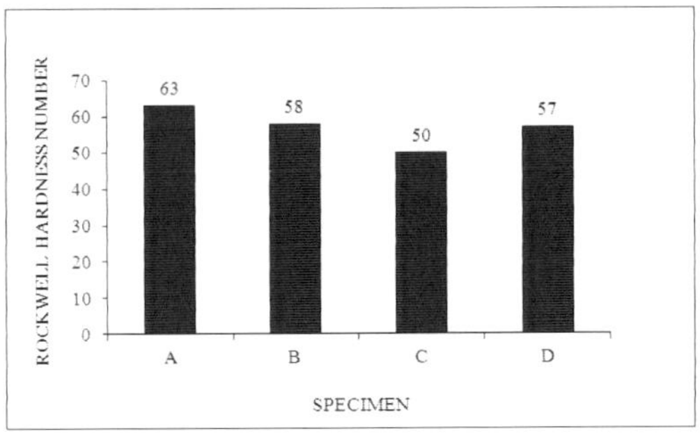


Fig. 6. Comparison of hardness value

### *4.2 Specimen Vs UTS*

Fig 7 illustrates the plot of Ultimate Tensile Strength (UTS) in N/mm2 versus specimens combinations, it can be seen that the treated palm leaf stalk fiber/sandwiched glass fiber /filled coconut shell nano filler reinforced polymer composite (specimen A) yielded a maximum of 123 MPa.

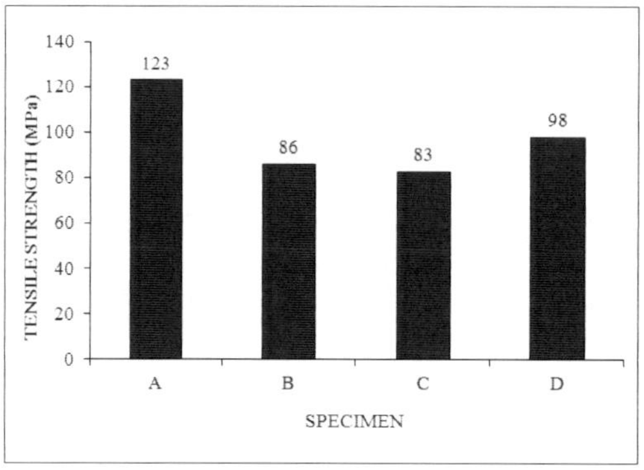


Fig. 7. Comparison of Ultimate Tensile Strength

*4.3 Specimen Vs Flexural strength*

The variation of flexural strength for all the tested specimens was shown in Fig 8. The specimen A recorded with increased flexural strength than the other specimens. This shows composite compatibility and fiber interaction with matrix.

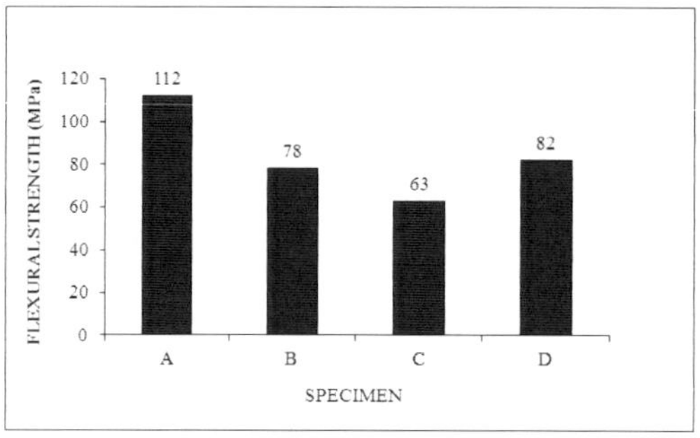


Fig. 8. Comparison of flexural strength

*4.4 Specimen Vs Impact strength*

The variation of impact strength for all the specimens was shown in Fig 9. The specimen A recorded increase in impact strength than the others which results in obtaining the enhanced impact properties.

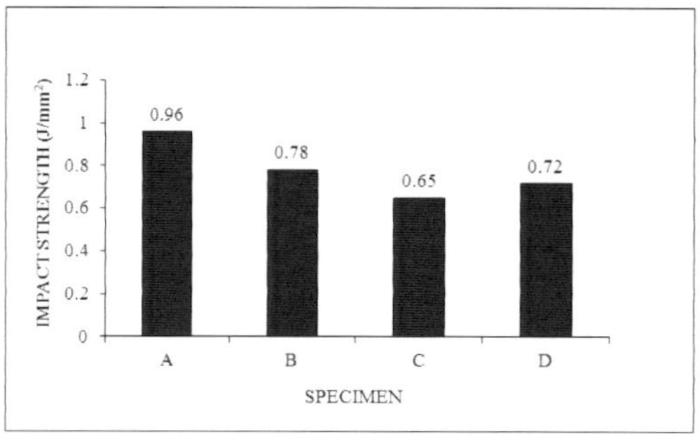


Fig. 9. Comparison of impact strength

## V. CONCLUSIONS

The fabrication of palm leaf stalk/glass sandwiched polymer composite was successfully completed. The fiber treatment by $KMnO_4$ soaking improved the mechanical properties like tensile strength, flexural strength, impact strength and hardness value. The various mechanical tests revel that mechanical properties of alkali treated fibers are superior to those of other compared natural fibers. Alkali soaking significantly removed surface impurities from fibers so a better fiber matrix adhesion was created. The result reveals that, the trend in all the properties increased for alkali treated and coconut shell nano powder added palm/glass fibers sandwich composites.

## REFERENCES

[1] Nabinejad. D, Sujan, M.E. Rahman, Ian J. Davies, "Effect of oil palm shell powder on the mechanical performance and thermal stability of polyester composites," Materials & Design. 2015; 65:823-830, 2015.

[2] M. Jawaid , H.P.S. Abdul Khalil, Azman Hassan, Rudi Dungani, A. Hadiyane. "Effect-of-jute-fibre-loading-on-tensile-and-dynamic-mechanical-properties-of-oil-palm-epoxy-composites." Composites-Part-B-Engineering.; 45:619-624, 2013.

[3] Y. Li and Y.O. Shen, "The use of sisal and henequen fibres as reinforcements in composites. Composite Materials,"; 6: 165-210, 2015.

[4] Parnia Zakikhani, R. Zahari, M.T.H. Sultan, D.L. Majid, "Extraction and preparation of bamboo fibre-reinforced composites. Materials & Design,"; 63:820-828, 2014.

[5] Emanuel M. Fernandes, Vitor M. Correlo. João F. Mano. Rui L. Reis. "Cork–polymer biocomposites: Mechanical structural and thermal properties," Materials & Design. 2015.

[6] Hind Abdellaoui, Hala Bensalah. Jamal Echaabi. Rachid Bouhfid. Abouelkacem Qaiss, "Fabrication characterization and modelling of laminated composites based on woven jute fibres reinforced epoxy resin. Materials & Design," 68:104-113, 2014.

[7] Elias Nassiopoulos, James Njuguna, "Thermo-mechanical performance of poly(lactic acid)/flax fibre-reinforced biocomposites. Materials & Design," 66: 473-485, 2015.

[8] Yan Li. Hao Ma. Yiou Shen. Qian Li. Zhuoyuan Zheng, "Effects of resin inside fiber lumen on the mechanical properties of sisal fiber reinforced composites," Composites Science and Technology,; 108:32-40, 2015.

[9] Pakanita Muensri, Thiranan Kunanopparat, Paul Menut, Suwit Siriwattanayotin, "Effect of lignin removal on the properties of coconut coir fiber/wheat gluten biocomposite Composites Part A: Applied Science and Manufacturing,"; 42:173-179, 2011.

[10] Xiangming Zhou. Seyed Hamidreza Ghaffar. Wei Dong. Olayinka Oladiran. Mizi Fan. "Fracture and impact properties of short discrete jute fibre-reinforced cementitious composites. Materials & Design"; 49:35-47, 2013.

[11] H.P.S. Abdul Khalil. M.S. Alwani. M.N. Islam. S.S. Suhaily. R. Dungani. Y.M. H'ng and M. Jawaid. "The use of bamboo fibres as reinforcements in composites, Composite Materials".; 16: 488-524, 2015.

[12] P. Noorunnisa Khanam. M.A. AlMaadeed. "Improvement of ternary recycled polymer blend reinforced with date palm fibre," Materials & Design.; 60:532-539, 2014.

[13] V. Fiore. T. Scalici. G. Di Bella. A. Valenza. "A review on basalt fibre and its composites, Composites Part B: Engineering."; 74:74-94, 2015.

[14] A. Alavudeen. N. Rajini. S. Karthikeyan. M. Thiruchitrambalam. N. Venkateshwaren. "Mechanical properties of banana/kenaf fiber-reinforced hybrid polyester composites: Effect of woven fabric and random orientation", Materials & Design.; 66:246-257, 2015.

[15] R. Yahaya. S.M. Sapuan. M. Jawaid. Z. Leman. E.S. Zainudin. "Effect of layering sequence and chemical treatment on the mechanical properties of woven kenaf–aramid hybrid laminated composites." Materials & Design.; 67:173-179, 2015.

[16] A. Hakamy. F.U.A. Shaikh. I.M. Low. "Thermal and mechanical properties of NaOH treated hemp fabric and calcined nanoclay-reinforced cement nano composites." Materials & Design.; 80:70-81, 2015.

[17] J. Sarki. S.B. Hassan. V.S. Aigbodion. J.E. Oghenevweta. "Potential of using coconut shell particle fillers in eco-composite materials." Journal of Alloys and Compounds; 509:2381-2385, 2011.

[18] BSJ Retnam, M Sivapragash, P Pradeep. "Effect of Fiber Orientation and Filler Material on Tensile and Impact Properties of Natural FRP", European Journal of Scientific Research.; 99:55-60, 2013.

[19] S.A.S. Goulart. T.A. Oliveira. A. Teixeira. P.C. Miléo. D.R. Mulinari. "Mechanical Behaviour of Polypropylene Reinforced Palm Fibers Composites." Procedia Engineering. ; 10:2034-2039, 2011.

[20] G.Velmurugan1, D.Vadivel, R.Arravind, S.P.Vengatesan, A.Mathiazhagan. "Tensile Test Analysis of Natural Fiber Reinforced composite." International Journal of Mechanical and Industrial Engineering. (IJMIE); 2,2012.

[21] N. W. A. Razak, A. Kalam, . "Effect of OPEFB Size on the Mechanical Properties and Water Absorption Behaviour of OPEFB/PPnanoclay/PP Hybrid Composites." Procedia Engineering; 41:1593 – 1599, 2012.

[22] S.M. Sapuana, D. Bachtiara, "Mechanical Properties of Sugar Palm Fibre Reinforced High Impact Polystyrene Composites." Procedia Chemistry.; 4:101-106, 2012.

[23] M.N.K. Chowdhurya, M.D.H. Bega, Maksudur R. Khana. "Biodegradability of nanoparticle modified fiber reinforced polyester resin nanocomposite." Procedia Engineering; 68:431 – 438, 2013.

[24] K.K. Ikpambese, D.T. Gundu, L.T. Tuleun, "Evaluation of palm kernel fibers (PKFs) for production of asbestos-free automotive brake pads" Journal of King Saud University - Engineering Sciences. 2014.

[25] Nadendla Srinivasababu. J. Suresh Kumar. K. Vijaya Kumar Reddy. "Manufacturing and Characterization of Long Palmyra Palm/Borassus Flabellifer Petiole Fibre Reinforced Polyester Composites." Procedia Technology. ; 14:252-259, 2014.

[26] A.Shalwan, B.F. Yousif . "Influence of date palm fibre and graphite filler on mechanical and wear characteristicsof epoxy composites." Materials and Design 2014; 59: 264–273, 2014.

# Influence of MgO particles on Microstructural and Mechanical Behaviour of AA7068 Metal Matrix Composites

K. John Joshua*
Research Scholar
Department of Mechanical Engineering,
Karunya University,
Coimbatore-641114, Tamil Nadu, India.
kjohnjoshua@gmail.com

S.J. Vijay, D. Philip Selvaraj, P.Ramkumar
Department of Mechanical Engineering,
Karunya University,
Coimbatore-6411114,
Tamil Nadu, India.

***Abstract***

Present work deals with the fabrication of AA7068 metal matrix composites reinforced with different weight percentages of MgO (0%, 1%, 2%, and 5%) produced using powder metallurgy route. A low pressure of 318 MPa was applied for compacting the composites and sintered at a temperature of 560°C for one hour. Optical microscopy was done to study the microstructural behavior. Wear test has been conducted for a sliding distance of 3 km with a sliding velocity of 1.2 m/s of load 5N.,7.5N and 10N. Vickers micro hardness test has been conducted and the maximum value of 68 VHN was obtained by adding a weight percentage of 5% MgO particles. The wear resistance has been improved by adding MgO particles in the matrix material. EDAX analysis was done to ensure the presence of MgO particles.

***Keywords: compacting, sintering, powder metallurgy, SEM***

## I. INTRODUCTION

Aluminum matrix composites are extensively studied due to their high wear resistance, good thermal conductivity, strength and high elastic modulus, relatively low density and functionality in high temperatures [1]. AA7068 is a strongest aluminum alloy having magnesium and zinc as its major alloying elements which has better mechanical properties compared to other Aluminum alloys. It offers high strength, good workability, good machinability, high resistance to corrosion and so on [2]. Among the 7xxx series, AA7068 is famous aluminum alloy, which provides higher mechanical strength. Higher strength is required to replace heavier metals with this alloy in some applications at ambient and elevated temperatures [3]. Power metallurgy provides better bonding of matrix and nanoparticles. MgO due its high melting point (Tm = 2800 _C), compressive strength, hardness and excellent thermodynamic stability is a better choice for reinforcement. The mechanical properties has been improved by the addition of MgO particles [4]. The hardness has been increased due to the addition of fine MgO particles into Zirconia Toughened Alumina (ZTA) and wear performance has been improved to 50% [5]. Increased wear loss and porosity was found with increased MgO reinforcement volume fraction [6]. Under 2N load, MgO coating showed an excellent resistance to sliding wear [7]. With ZTA, Vickers hardness increases with finer MgO particle size and gradually decreases with coarser MgO particle size [8]. The increases in magnesium oxide content up to 5% led to a decrease in the shrinkage and density [9]. With the increase of MgO the hardness of the composite specimens increases and the amount of resistance decreases [10].

In this paper, microstructural and mechanical properties of AA7068 reinforced with MgO particles, which were manufactured using powder metallurgy were reported.

* Corresponding author: mail id: kjohnjoshua@gmail.com
Phone: 91-9894341750

IOP Conf. Series: Materials Science and Engineering **247** (2017) 012011 doi:10.1088/1757-899X/247/1/012011

## II. EXPERIMENTAL PROCEDURE

### *A. Preparation of Composites*

The AA7068/MgO composites were produced using powder metallurgy. Table 1 shows the composition of AA7068 powders that were weighed accurately and mixed in a ball mill for 40 hours in a high energy ball mill [3]. 2% stearic acid was added to the powders to have a strong bonding [11]. A separate die and punch was made for compacting the metal powders. Cold compaction at a low pressure of 318 MPa was done using a Universal Testing Machine (UTM) to produce green compacts of size 20 mm diameter and 32±2 mm height. The green compacts were sintered at 560°C for one hour in a sintering furnace (fig 4). The composites of AA7068 reinforced with MgO (0%, 1%, 2%, 5%) were produced which were shown in Fig. 3. Fig. 1 and fig 2 shows the SEM images of AA7068 and MgO powders respectively. Table II shows the average particle size of the powders analyzed from particle size analyzer.

TABLE I. COMPOSITION OF AA7068 (% BY WEIGHT)

| Element | Si | Fe | Cu | Mn | Mg | Cr | Zn | Ti | Zr | Al |
|---|---|---|---|---|---|---|---|---|---|---|
| Content | 0.12 | 0.15 | 2 | 0.1 | 3 | 0.05 | 8 | 0.01 | 0.1 | Bal |

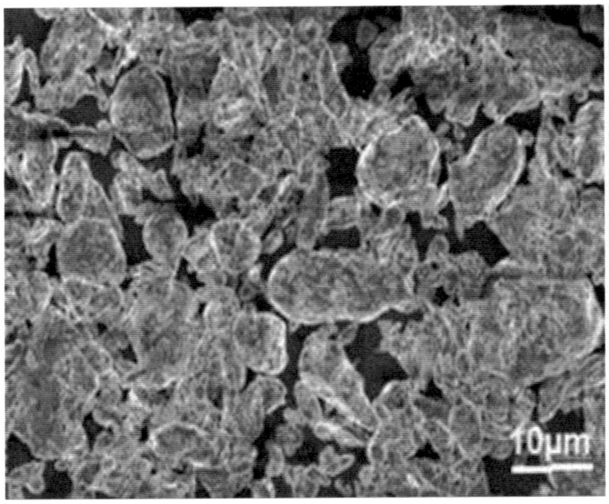


Fig.1 SEM image of AA7068 powders

Fig. 2. SEM image of MgO powders

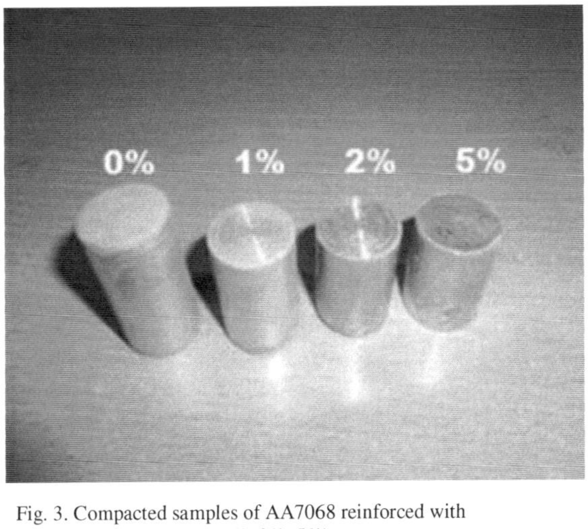


Fig. 3. Compacted samples of AA7068 reinforced with MgO (0%, 1%, 2%, 5%)

Fig. 4. Sintering furnace

Fig. 5. Pin-on-disc wear testing machine

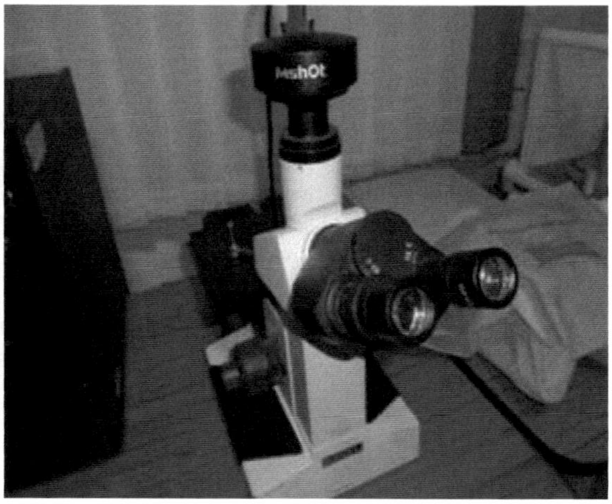

Fig. 6. Vickers microhardness tester

### *B. Hardness Test*

The hardness were measured using Vickers micro-hardness tester. The test was conducted on polished sample of AA7068 and its composites. The test was carried at ten different locations and the averages were reported in table III. The Vickers microhardness tester was shown in the Figure 6.

TABLE II. AVERAGE PARTICLE SIZE

| Powder | Average Particle size (μm) |
|---|---|
| AA7068 | 1.785 |
| MgO | 3.619 |

### *C. Wear Test*

The sintered samples were extruded to10 mm diameter and 30 mm height for testing purpose. Wear tests were carried out using DUCOM TR 20-LE pin-on-disc wear testing machine. A sample size of 10 mm diameter and 30 mm height was used for testing. Initially the pin masses were weighed using Shimadzu electronic weighing machine ATY 224 model with an accuracy of 0.0001g. All tests were conducted at room temperature (26$^{o}$ C). The experiments were conducted at a load of 5N, 7.5N, 10N at a sliding velocity of 1.2 m/s at six sliding distances for a total of 3 km. The exact mass was measured after every sliding distance of 500 m after the wear test.

## III. RESULTS AND DISCUSSION

### *A. Microstructural analysis*

Fig. 3 shows the samples after cold compaction. After compaction the samples were sintered at 560$^{o}$C in a sintering furnace for 1 hour. The samples were polished and grinded before conducting the optical microscopy. The surface of the samples were ground by using 600,800 and 1200 grit abrasive paper and polished using diamond paste. Fig. 7 shows the optical micrographs of AA7068 reinforced with MgO at (a) 0% (b) 1% (c) 2% and (d) 5%. MgO particles were uniformly distributed in the matrix. Fig. 7(a) shows more porosity present in the AA7068 sample without reinforcement, with more number of fine and spherical shape of pores. This is due to the application of low applied pressure (318 MPa) which coincides with Amin et al [3]. Fig. 7(b) shows excellent bonding between the AA7068 particles and MgO particles with very less pore holes and a very high density. This is due to fine particles of MgO mixed with the matrix material. Smaller the grain size improves the hardness, toughness and reduces wear rate [12]. Fig. 7(c) shows the particles are uniformly distributed, but having few non-uniform larger cavities which is due to poor pressing which goes parallel with Amin et al [3]. Fig. 7(d) shows that the particles with homogeneous distribution, but few non-uniform large shaped pores due to poor pressing. Porosity is more because aggregation of MgO nanoparticles is less probable in powder metallurgy. Formation of porosity is highly probable in sintering method compared to other methods [4].

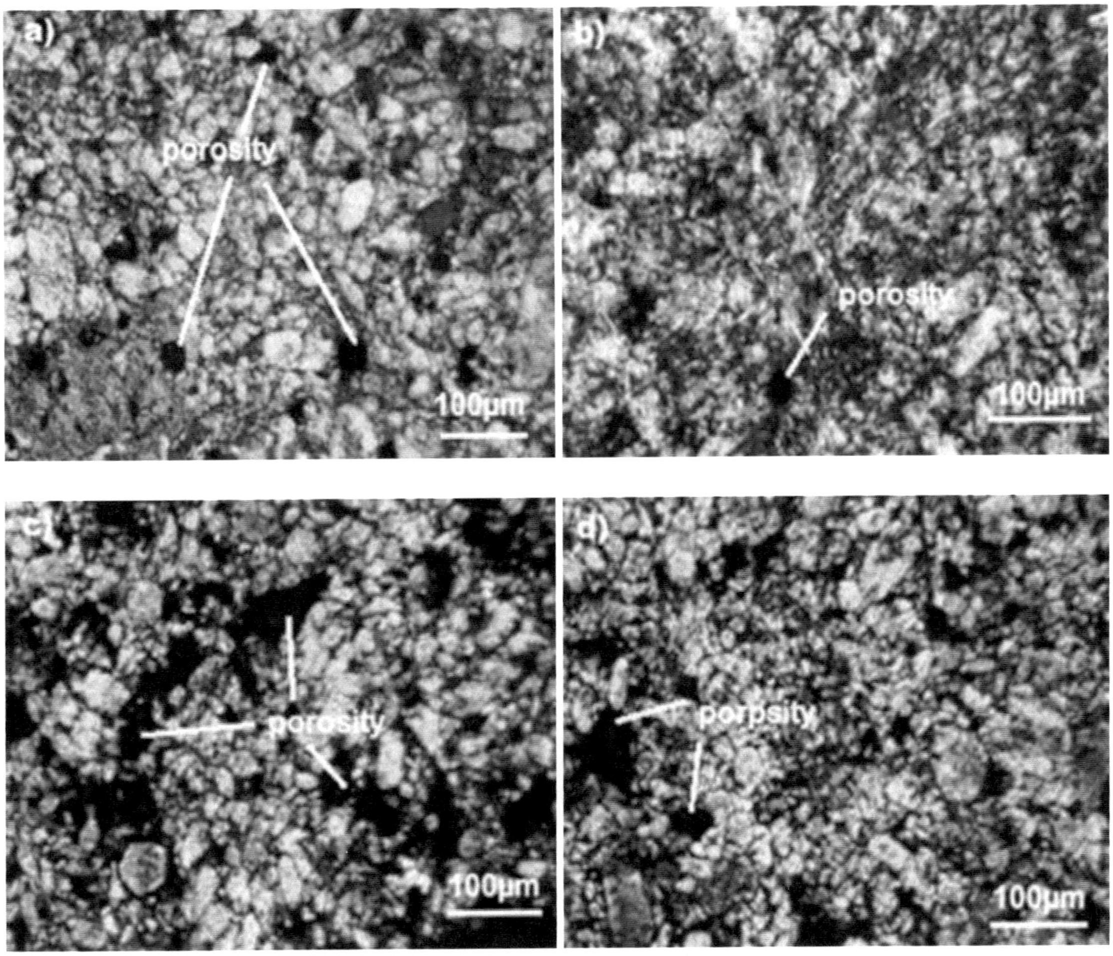


Fig. 7: Optical Micrographs of AA7068 reinforced with MgO (a) 0% (b) 1% (c) 2% and (d) 5%

### *B. Wear behavior*

Fig 8. shows the graph of sliding distance against wear rate plotted for AA7068/MgO composites. The wear rate increases with increasing sliding distance, while as the MgO weight fraction increases, wear rate decreases. The wear rate was found to be maximum for AA7068 without reinforcement and found to be minimum for the maximum percentage of the addition of MgO particles. As the load increases, wear rate also increases. The wear rate for 10 N load is more compared to 7.5 N load and similarly wear rate for 7.5 N load is more compared to 5 N load.

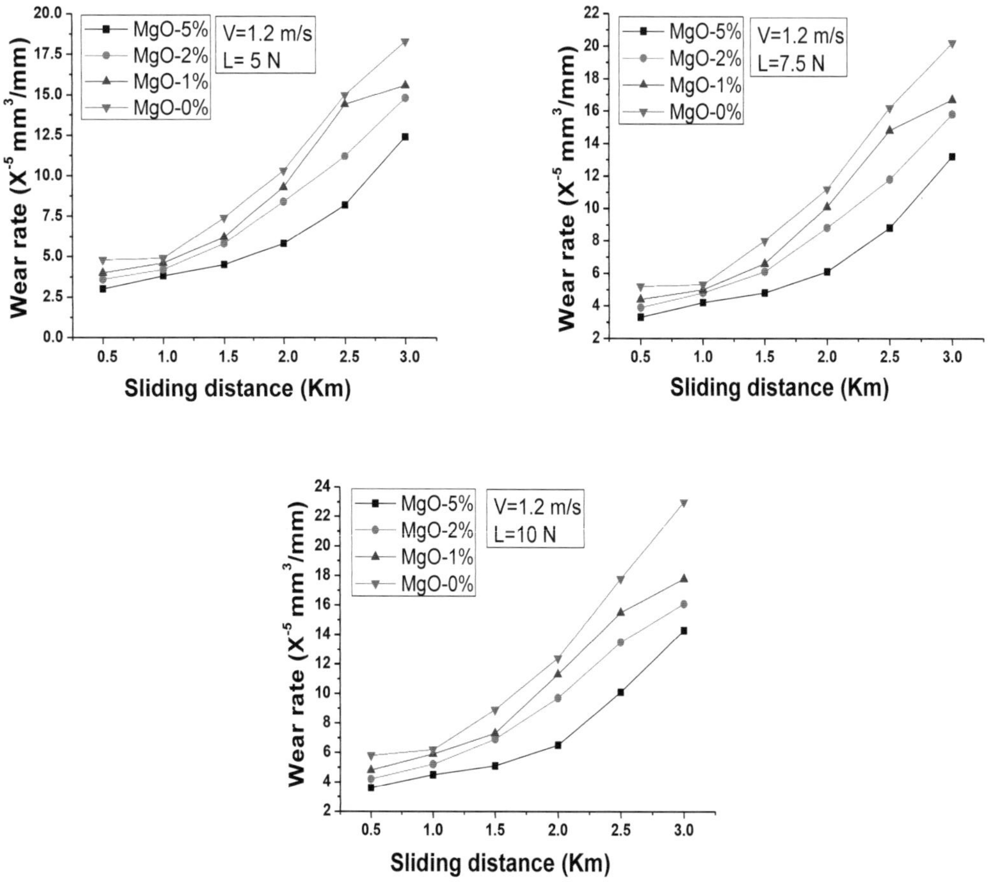


Fig 8. Wear rate Vs sliding distance of AA7068 reinforced with MgO particles

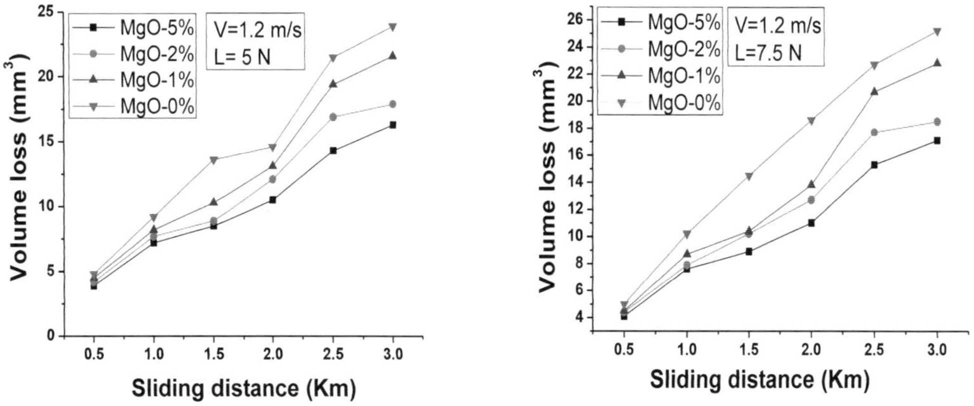

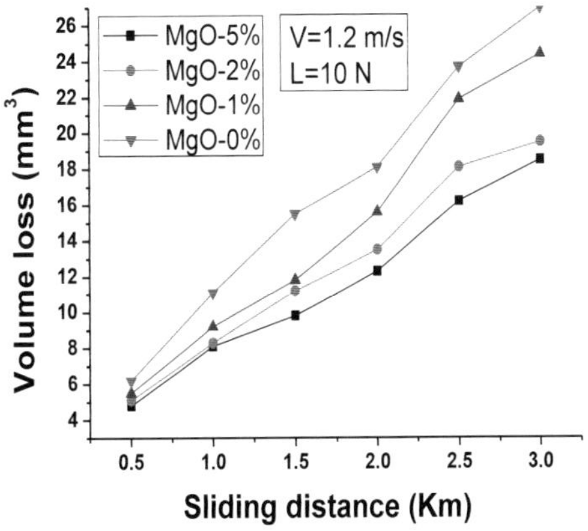


Fig 9. Volume loss Vs sliding distance of AA7068 reinforced with MgO particle

Fig 9. shows the graph of sliding distance against volume loss plotted for AA7068/MgO composites. The wear curves shows that the volume losses of all produced composites decreases with increasing MgO weight fraction and increases with increasing sliding distance. The volume loss was found to be maximum for AA7068 without reinforcement and found to be minimum for the maximum percentage of addition of MgO particles. As the load increases, the volume loss also increases. The volume loss for 10N load is maximum compared to the volume loss for load of 5N and 7.5N.

There is a good agreement between the hardness values given in table III and the wear behavior of the composites. The wear resistance of the composites increases with increase in their hardness values.

### *C. EDAX*

Fig. 10. Illustrates the EDAX pattern for AA7068 reinforced with MgO a) 0% b) 1% c) 2% and d) 5%. The EDAX pattern in fig. 10 (b), 6(c) and 6(d) confirms the presence of magnesium oxide particles.

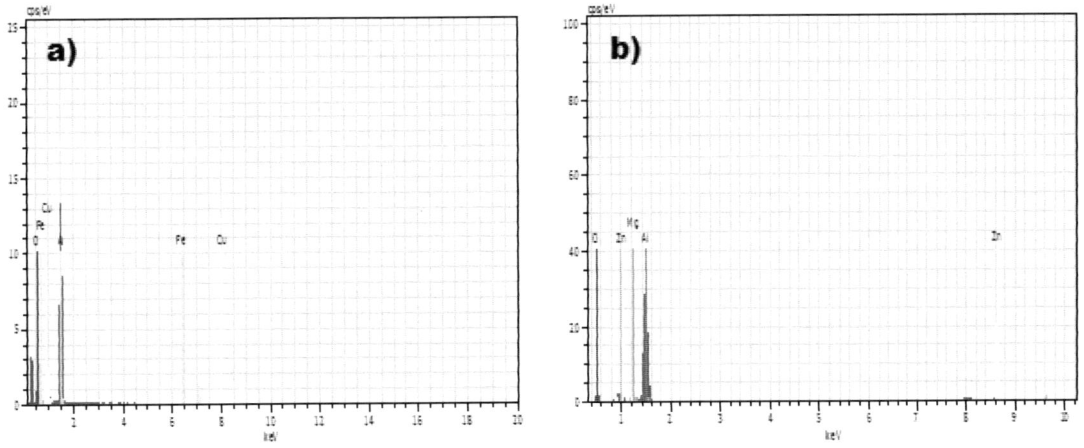

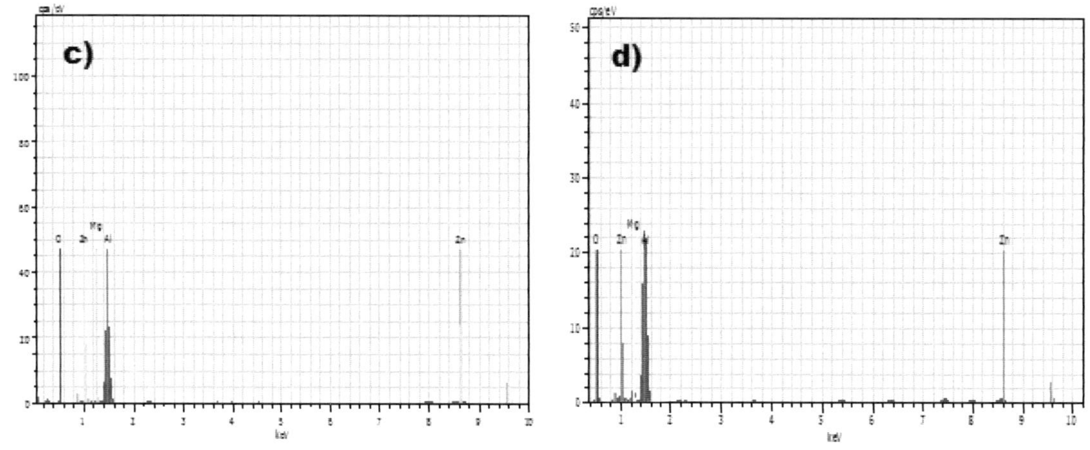


Fig 10. EDAX pattern for AA7068 reinforced with MgO a) 0% b) 1% c) 2% and d) 5%

### *E. Hardness Test*

The hardness test was conducted using Vickers micro hardness tester. The hardness values of AA7068/MgO composites were given in the table III.

TABLE III. VICKERS HARDNESS NUMBER

| Composite | Hardness(VHN) |
|---|---|
| AA7068 | 33 |
| AA7068 + 1% MgO | 48 |
| AA7068 + 2% MgO | 56 |
| AA7068 + 5% MgO | 68 |

As the weight fraction of MgO increases, the hardness also increases and was found to be a maximum of 68 for 5% MgO, because of the homogeneous distribution of MgO particles and due to the decrease in grain size of the particles [12].

## IV. CONCLUSION

1. AA7068/MgO composites can be produced successfully using powder metallurgy.
2. The optical micrographs showed that the particles were mixed with homogeneous dispersion.
3. The Vickers microhardness number has been increased to a maximum of 68 VHN for an addition of 5% MgO particles.
4. The wear resistance has been improved by adding MgO particles in AA7068 matrix material.

## ACKNOWLEDGMENT

The authors would like to thank Mr. J. Devamanoharan of Centre for Research in Metallurgy laboratory of Karunya University for providing his valuable guidance and support.

## REFERENCES

[1] Yuhai Dou, Yong Liu, Yanbin Liu, Zhiping Xiong, Qingbing Xia,"Friction and wear behaviors of B4C 6061Al composite" , Materials and Design 60 (2014) 669–677.
[2] Madhusudhan M, Naveen GJ, Mahesha K, "Mechanical Characterization of AA7068-ZrO2 reinforced Metal Matrix Composites"*Materials Today: Proceedings,* 2017,4, 3122–3130.

[3] Amin AZIMI, Ali SHOKUHFAR, Omid NEJADSEYFI, Hamid FALLAHDOOST, and Saeid SALEHI, "Optimizing consolidation behavior of Al 7068–TiC nanocomposites using Taguchi statistical analysis", Trans. Nonferrous Met. Soc. China 25(2015) 2499–2508.

[4] Ahmad Zahirani Ahmad Azhara, Hasmaliza Mohamada, Mani Maran Ratnamb, Zainal Arifin Ahmad,"The effects of MgO addition on microstructure, mechanical properties and wear performance of zirconia-toghned alumina cutting inserts", Journal of Alloys and Compounds 497(2010) 316-320.

[5] Hossein Abdizadeh, Reza Ebrahimifard, and Mohammad Amin Baghchesara,"Investigation of microstructure and mechanical properties of nano MgO reinforced Al composites manufactured by stir casting and powder metallurgy methods A comparative study", Composites: Part B 56(2014) 217-221.

[6] Muharr em Pul, Recep Çalin, Ferhat Gül,"Investigation of abrasion in Al-MgO metal matrix composites", Materials Research Bulletin, Volume 60, December 2014, Pages 634-639.

[7] P. Bala Srinivasan, J. Liang, C. Blawert, W. Dietzel,"Dry sliding wear behaviour of magnesium oxide and zirconium oxide plasma electrolytic oxidation coated magnesium alloy", Applied Surface Science, Volume 256, Issue 10, 1 March 2010, pp 3265-3273.

[8] Ahmad Zahirani Ahmad Azhar, Hasmaliza Mohamad, Mani Maran Ratnam, Zainal Arifin Ahmad, "Effect of MgO particle size on the microstructure, mechanical properties and wear performance of ZTA–MgO ceramic cutting inserts", Int. Journal of Refractory Metals and Hard Materials 29 (2011) 456-461.

[9] V. Sirota, O. Lukianova, V. Krasilnikov, V. Selemenev, V. Dokalov," Microstructural and physical properties of magnesium oxide-doped silicon nitride ceramics", Results in Physics 6 (2016) 82–83.

[10] Muharr em Pul, Recep Çalin, Ferhat Gül,"Investigation of abrasion in Al–MgO metal matrix composites", Materials Research Bulletin 60 (2014) 634–639.

[11] P.C.Angelo and R.Subramanian, "Powder Metallurgy: Science, Technology and Application", Prentice Hall,India, 2008.

[12] T. Alamb, K.H.A. Al Mahmuda, M.F. Hasanb, S.A. Shahira, H.H. Masjukia, T. Alamb, K.H.A. Al Mahmuda, M.F. Hasanb, S.A. Shahira, H.H. Masjukia, M.A. Kalama, A. Imrana, H.M. Mobaraka, "Effect of Magnesia on Zinc Oxide Stabilized Nano Alumina Ceramic powder in sintering process and its impact on mechanical properties", Procedia Engineering 68 ( 2013 ) 723 – 729.

[13] Ali Alizadeh, Alireza Abdollahi, Hootan Biukani, "Creep behavior and wear resistance of Al 5083 based hybrid composites reinforced with carbon nanotubes (CNTs) and boron carbide (B4C)", Journal of Alloys and Compounds 650 (2015) 783-793.

# *Investigations on LM6 Metal Matrix Composite with borosilicate Glass Reinforcement for Aerospace applications*

J.David Rathnaraj**
Department of Aeronautical Engineering
Sri Ramakrishna Engineering College
Coimbatore, India
davidrathnaraj@gmail.com

S.Sathish
Department of Mechanical Engineering
Sri Ramakrishna Engineering College
Coimbatore, India
sathishresearch2012@gmail.com

*Abstract*—**The recycling of glass wastes from the industries and society holds a threat to the environment and leads to the need for new applications. While producing a metal matrix composite production cost is an important factor which decides the suitable application. So, while developing a new material with this low – cost has great importance in this competitive world. In this study, an metal–matrix composite fabricated from an aluminum alloy (LM6) and Borosilicate glass powder particles with % addition of 2.5%, 5%, 7.5%, and 10% were produced by liquid Processing (stir casting) technique. The variations in the mechanical properties like toughness, compressive strength, hardness, and tensile were examined. The microstructures of the fabricated metal matrix composite have been obtained by using Metallographic microscope. The addition of the borosilicate glass indicated an improved behavior in the hardness and toughness properties. The Rockwell hardness value of fabricated metal matrix composite increases with the increase in % of reinforcement. The compressive and tensile strength of the fabricated MMC increases until reinforcement reaches a maximum of 7.5%. The microstructure of the fabricated MMC shows that the reinforcements were homogeneously distributed in the fabricated metal matrix composite.**

***Keywords—Borosilicate Glass; LM-6; metal matrix composites; Heat treatment;***

## I. INTRODUCTION

Metal matrix composites were used in the fields of automotive and aeronautical applications due to their attractive properties compared to monolithic materials [8]. Applications of aluminium alloys in various fields have been increased due to its light weight material and good corrosion resistance properties. The aluminium matrix composites have emerged from the lower mechanical properties of the pure aluminum. The high strength, and light weight made the AMC desirable in many applications, mainly on automobile products such as engine piston, cylinder liner, brake disc/drum etc [6]. Also, the ceramic reinforcement will create much strong and hard phases in aluminium matrix which will give good wear resistance compared to pure metals [15]. Insitu process showed better results for manufacturing PMMC's Impact of $Zr\ B_2$ reinforcement particles in Al 6061 were discussed [14]

While producing a metal matrix composite cost is the key factor in their wider application in various fields. It mainly depends on two factors that are the type of reinforcement using and the technique which is being used to produce the MMC. Only simpler fabrication methods, higher production volumes, and use of cheaper reinforcements can reduce the cost of MMCs [5]

It is found that most of the aluminium metal matrix composite uses reinforcements such as SiC, TiC, $Al_2O_3$ which are costlier and have a higher density [10]. This will increase the total weight of the components. The stir casting method improves the homogeneous dispersion of the glass particulates and it will provide a good surface finish and better cooling rate [3].

## II. MATERIALS

*A. Matrix material*First, LM6 alloy is a eutectic alloy which is having a major composition of 85.95% of aluminium and 11% to 13% of silicon. Its ability to resist hot cracking, pressure tightness, die-filling capacity, and corrosion resistance makes it as a commonly used alloy for automotive as well as the aeronautical applications [2]. LM-6 has an excellent cast ability and fluidity which permits the alloy to be used of intricate and complex castings. It is best for service operating conditions require good resistance to corrosion it offers better corrosion resistance than LM-2 and LM-24. The chemical composition of LM6 aluminium alloy used is given in table 1.

Table1: Chemical Composition of LM6

| Material | Percentage |
|---|---|
| Copper | 0.1 max. |
| Magnesium | 0.10 max. |
| Silicon | 10.0-13.0 |
| Iron | 0.6 max |
| Manganese | 0.5 max |
| Nickel | 0.1 max. |
| Zinc | 0.1 max. |
| Lead | 0.1 max. |
| Aluminum | Remainder |
| Titanium | 0.2 max |

*B. Matrix material*

When producing a metal matrix composite cost is a main factor for their wider application in automotive and aerospace industries. Glass is a low-cost reinforcement material which easily available and will significantly reduce the cost of MMCs, which make this study a significant one. The borosilicate glass has good chemical and thermal resistance and is mainly used for making modern laboratory glassware. The lesser co-efficient of thermal expansion also provides a better bonding between aluminium matrix. The chemical composition of borosilicate glass is shown in table 2

Table 2: Chemical Composition of Borosilicate glass

| Material | Percentage |
|---|---|
| $SiO_2$ | 80.6 |
| $B_2O_3$ | 13.0 |
| $Na_2O$ | 4.0 |
| $Al_2O_3$ | 2.3 |

## III. EXPERIMENTAL PROCEDURE

The borosilicate glass from a recycling plant was taken and cleaned by using water. The dried glass sample was crushed with the help of a ball mill and seived to the size of 105μm. 1Kg of commercially available LM 6 and desired amount of Glass particles were taken in the preparation of each sample. The Glass particle was preheated to 400°C for 30 min to remove moisture. LM 6 was melted in a resistance furnace. The melt temperature was raised up to 850°C. Small amount of magnesium was added to the molten metal to increase the wettability of the aluminium melt. The melt was stirred with the help of a stainless steel stirrer. The stirring was maintained between 5 min at an impeller speed 300 rpm.

*A. Tensile testing*

The tensile tests were conducted on the UTM at room temperature. The samples were prepared as per the standard ASTM B557. The test has been performed to find out the ultimate strength of the composite material. The results from the test are given in table 4.

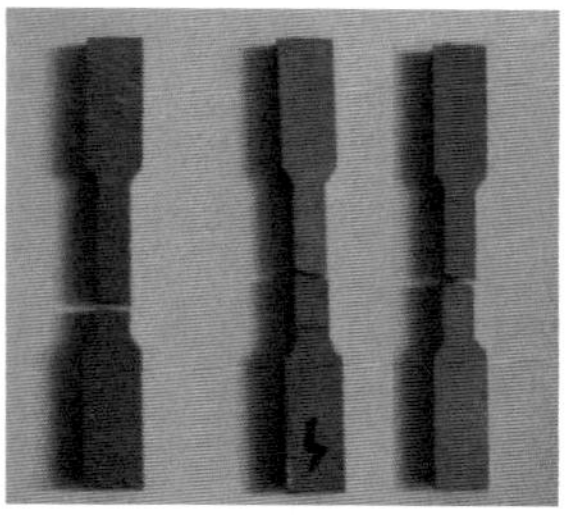

**Figure 1: Tensile test specimens**

*B. Impact test*

The impact tests were conducted at room temperature and the specimens were made according to ASTM standard E23. Both Izod and Charpy impact test were performed on the borosilicate glass reinforced samples.

**Figure 2: Impact test specimens**

*C. Rockwell hardness test*

The hardness test was conducted in the Rockwell hardness testing machine. Hardness Specimen of dia 20mm and length 20 mm were cut from the cast samples and the surface were ground using a grinding machine. The indenter has 1.588 mm diameter and a force of 100Kgf was applied on the surface for 30 Sec.

*D. Compression test*

The specimens for compression test were prepared according to ASTM standard E9–89a. The test was performed by using Universal Testing Machine (UTM) at room temperature. The ultimate compression strength that it can withstand ware found out.

## IV. RESULTS AND DISCUSSION

*A. Density analysis*

The densities of the MMC were determined by using the Archimedes principle. The samples were weighed by using a digital weighing scale. The volumes of the samples were measured by the rise in water level. The results are tabulated in table 3.

Table 3: Density of the composite material

| No | % Reinforcement of Glass | Actual Density ($g/cm^3$) |
|---|---|---|
| 1 | 0 | 2.68 |

| | | |
|---|---|---|
| 2 | 2.5 | 2.66 |
| 3 | 5.0 | 2.65 |
| 4 | 7.5 | 2.636 |
| 5 | 10 | 2.627 |

The result shows that the densities of the sample were decreasing as the % reinforcement increases. This will helps to reduce the overall weight of the component.

### *B. Tensile test*

The tensile test results are shown in the table 4. From the results it is found that the LM6 is having a tensile strength of 105.185 MPa at room temperature.

Table 4: Tensile test results

| No | % Reinforcement of Glass | Tensile strength (MPa) |
|---|---|---|
| 1 | 0 | 105.185 |
| 2 | 2.5 | 112.270 |
| 3 | 5.0 | 126.712 |
| 4 | 7.5 | 147.695 |
| 5 | 10 | 135.160 |

The tensile strength tends to increase as the % reinforcement increases and reaches a maximum of 147 MPa at 7.5 % of reinforcement and then it started to decrease due to the clustering borosilicate glass which permits crack propagation.

### *C. Rockwell Hardness test*

The hardness test results were shown in the table 5. It shows that the Rockwell hardness of pure LM6 has 51 HRB at room temperature. The hardness of the metal matrix composite increases as the %reinforcement increases and achieved a maximum value of 70 HRB.

Table 5: Rockwell Hardness the composite material

| No | % Reinforcement of Glass | Trial | | | HRB |
|---|---|---|---|---|---|
| | | 1 | 2 | 3 | |
| 1 | 0 | 51 | 51 | 51 | 51 |
| 2 | 2.5 | 56 | 56 | 56 | 56 |
| 3 | 5.0 | 62 | 62 | 62 | 62 |
| 4 | 7.5 | 67 | 67 | 67 | 67 |
| 5 | 10 | 70 | 70 | 70 | 70 |

*D. Impact test*

Both Izod and Charpy impact tests were conducted on the borosilicate glass reinforced specimens at room temperature.
The impact test results were shown in the table 6 and 7. The results shows by adding borosilicate glass the impact strength increases from the pure LM6 and get a maximum of 4 J in charpy and 4.7 J in Izod.

Table 6: Charpy impact test results

| No | % Reinforcement of Glass | Impact energy (J) |
|---|---|---|
| 1 | 0 | 1.5 |
| 2 | 2.5 | 2.3 |
| 3 | 5.0 | 2.6 |
| 4 | 7.5 | 3.4 |
| 5 | 10 | 4 |

Table 7: Izod impact test results

| No | % Reinforcement of Glass | Impact energy (J) |
|---|---|---|
| 1 | 0 | 3 |
| 2 | 2.5 | 3.4 |
| 3 | 5.0 | 4 |
| 4 | 7.5 | 4.3 |
| 5 | 10 | 4.7 |

The results shows that impact strength increases as the % reinforcement increases.

IOP Conf. Series: Materials Science and Engineering **247** (2017) 012012 doi:10.1088/1757-899X/247/1/012012

*E. Compression test*

The Compression test results of the MMC are shown in the table 8. It shows the compression strength will increase by adding the borosilicate glass particulate in to the aluminium matrix at room temperature.

Table 8: Compression strength of the composite material

| No | % Reinforcement of Glass | Compression strength (MPa) |
|---|---|---|
| 1 | 0 | 266.069 |
| 2 | 2.5 | 303.024 |
| 3 | 5.0 | 369.541 |
| 4 | 7.5 | 384.323 |
| 5 | 10 | 376.932 |

The compression strength of the composite increases as the reinforcement increases up to 7.5 % and then reduces. The maximum strength that it can with stand was found out to be 384.323(MPa)

*F. Microstructure analysis*

The figures show the optical microstructure of different % of glass reinforcement. From the microstructure it is observed that blocky Si phase with generally circular shape is distributed in aluminum dendrites because of rapid solidification. The glass particles were homogeneously distributed throughout the matrix and good bonding between the aluminium and glass particles were observed. During solidification the glass particle are pushed by the aluminum dendrites into the last freezing eutectic liquid and thus the glass particles are seen surrounded by the eutectic silicon.

Figure 1: Optical Microstructure of LM 6 – 2.5% glass reinforcement

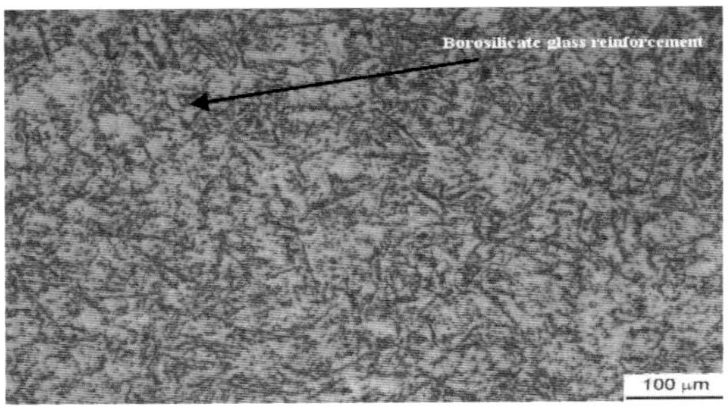


Figure 2: Optical Microstructure of LM 6 – 5.0% glass reinforcement

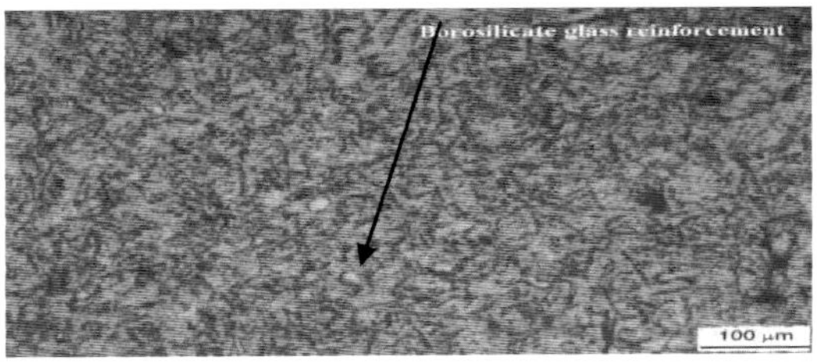

Figure 3: Optical Microstructure of LM 6 – 7.5% glass reinforcement

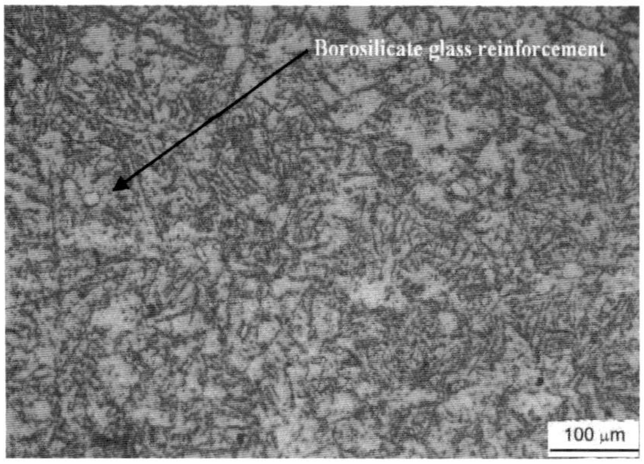


## V. CONCLUSION

- The uniform distribution of reinforcement of Borosilicate glass in LM6 alloy is confirmed through the microstructures. Hence MMC ingots can also be used for Automotive and aerospace industries.
- In the fabricated metal matrix composite materials As the percentage of reinforcement i.e. Borosilicate glass varies there is variation in the mechanical properties of like Hardness value, Impact energy, Tensile and compression strength.
- As the percentage of reinforcement increases i.e borosilicate glass the following observations were obtained in the impact energy there is a increases in charpy test in the range of 53.33%, 13.04%, 30.769% and 17.64%.
- As the percentage of reinforcement increases i.e. borosilicate glass the following observations were obtained in the Izod test impact energy gets consequently increases by 13.33%, 17.64%, 7.5% and 9.3%.
- While conducting the tensile test of metal matrix composite it increases up to 7.5% of borosilicate reinforcement while further addition shows an behavior of increasing pattern 6.735%, 12.86% and 16.55% and decreases up to 8.487%.
- Similarly for compressive strength of the MMC increases up to 7.5% of borosilicate reinforcement addition shows an behavior of increasing pattern 13.889%, 21.951% and 4.00% and then decreases upto 1.92%.Hence it is clear that addition of (Borosilicate glass) about 7.5%. In LM6 as reinforcement would provide better mechanical properties.

## REFERENCES

[1] Shamsul Baharin Jamaludina , Josef Hadipramana, Mohd Fitri Mohd Wahid, Kamarudin Hussin, Azmi Rahmat, "Microstructure and Interface Analysis of Glass Particulate Reinforced Aluminum Matrix Composite", Advanced Materials Research, 795 (2013), 578-581

[2] Sourav Kayala, R. Beherab, G.Sutradhara, "Mechanical properties of the as-cast silicon carbide particulate reinforced Aluminium alloy Metal Matrix Composites", International Journal of Current Engineering and Technology,2, (2012), 318-322

[3] Aondona P. Ihom, Nyior G. Bem, Emmanuel E. Anbua, Joy N. Ogbodo, "The Effect of Ageing Time on Some Mechanical Properties of Aluminum/0.5% Glass Reinforced Particulate Composite" Journal of Minerals and Materials Characterization and Engineering, 11 (2012), 919-923

[4] E. Bernardoa,, G. Scarincia, A. Maddalenaa, S. Hreglichb, "Development and mechanical properties of metal–particulate glass matrix composites from recycled glasses" Composites, 35 (2003), 17–22

[5] Madhu Kumar YC, Uma Shankar, "Evaluation of Mechanical Properties of Aluminum Alloy 6061-Glass Particulates reinforced Metal Matrix Composites", International Journal of Modern Engineering Research, 2, (2012), 3207-3209

[6] Nikilesh Chawla, "Metalmatrix in Automotive applications", Advanced materials & processes, 2006, 29-31

[7] Hu seyin Sevik, S. Can Kurnaz, "Properties of alumina particulate reinforced aluminum alloy produced by pressure die casting", Materials & Design 27 (2005)

[8] A.M.S.Hamouda, S.Sulaiman, T.R Vijayaram, M.Sayuti, M.H.M.Ahmad, "Processing and characterization of particulate reinforced aluminum silicon matrix composite" Journal of achievements in materials and manufacturing engineering, 25 (2007),

[9] S. Sulaiman, M. Sayuti, R. Samin, "Mechanical properties of the as-cast quartz particulate reinforced LM6 alloy matrix composites" journal of materials processing technology, 2008

[10] Suraya Sulaiman, Shamsuddin Sulaiman, Nur Najmiyah Jaafar, and Nor Imrah Yusoff, "Studies on Tensile Properties of Titanium Carbide (TiC) Particulates Composites"

[11] S.N. Aqida , M. I. Ghazali, J. Hashim, "The Effects of Stirring Speed and Reinforcement Particles on Porosity Formation in Cast MMC"

[12] G. G. Sozhamannan, S. Balasivanandha Prabu, V. S. K. Venkatagalapathy, "Effect of Processing Paramters on Metal Matrix Composites: Stir Casting Process" Journal of Surface Engineered Materials and Advanced Technology, 2, (2012)

[13] Azrol Jailania and Siti Mariam Tajuddin, "Mechanical Properties of Stirred SiC Reinforced Aluminium Alloy: Stir Casting With Different Composition of SiC, Blade Angle and Stirring Speed" Advanced Materials Research, (2013), pp 1335-1339

[14] I. Dinaharan, N. Murugan, Siva Parameswaran, "Influence of in situ formed ZrB2 particles on microstructure and mechanical properties of AA6061 metal matrix composites" Materials Science and Engineering A, (2011), 528 pp 5733–5740

[15] I. Dinaharan, "Influence of ceramic particulate type on microstructure and tensile strength of aluminum matrix composites produced using friction stirprocessing" Journal of Asian Ceramic Societies, (2016), 4 pp 209–218

# *Development of Fuzzy Logic Tool to predict Wear Rate in Aluminum Composite*

Rajesh Prabha[1], Edwin Raja Dhas.J[2]
[1]Department of Mechanical Engineering, [2]Department of Automobile Engineering
Noorul Islam University, Thuckalay, 629180
Tamil Nadu, India
[1]nrprabha@rediffmail.com, [2]edwinrajadhas@rediffmail.com

*Abstract*— **Industrial automation is the need of the hour. Success and failure of automation depends on the selection and efficient utilization of soft computing tools such as artificial neural network , expert systems, and fuzzy logic. This article explores the idea of constructing a fuzzy logic model to predict wear rate of fabricated composites under predetermined conditions of input variables such as applied load, sliding velocity and distance travelled. Taguchi design of experiments is employed for experimentation. 27 sets of experiments are performed by varying the parameters of applied load, sliding velocity and distance travelled. Experiment is repeated for average values of wear rate are measured to reduce experimental error. Matlab toolbox functions are used to build the model. Confirmatory experiments are done and the results are measured in terms of accuracy and time. It is found that the developed model predicts wear rate with acceptable limit. Hence the constructed model can be forwarded to predict wear rate of the developed composite under different conditions.**

***Keywords—Fuzzy Logic; Composite; Taguchi Method; wear rate;***

## I. Introduction

Recently, application of composite materials has been drastically increased in automotive, marine and aeronautical sectors due to light weight ratio and excellent mechanical properties. Commercial, defence and industrial firms use composite materials in an efficient way [1]. The guarantee of any product depends on wear rate. It is essential to check the tribological properties of the new developed materials (2,3). Better understanding of the relationship between the parameters involved, work materials and working conditions is an essential requirement for the prediction of wear rate (4). It is time consuming and tedious task to identify the relationship of the parameters involved. Mathematical modeling needs lot of time for computation. Hence the problem is solved by the usage of soft computing tools. Intelligent techniques such as fuzzy logic, ANN, and ANFIS can address this issue. Many authors have applied these techniques for wear rate prediction of the developed composite. ANN is applied to predict wear ( 5-8) mechanical (9) and tribological (10) properties. Radial basis functional neural network is applied to characterize the wear model of nano composite (11). This paper addresses the prediction of wear characteristics using fuzzy modeling.

## II. Proposed Methodology

The scheme of development of fuzzy model is shown in figure 1. The stages of development are a) Design and conduction of the experiment b) Development of fuzzy logic model c) Validation of the developed model.

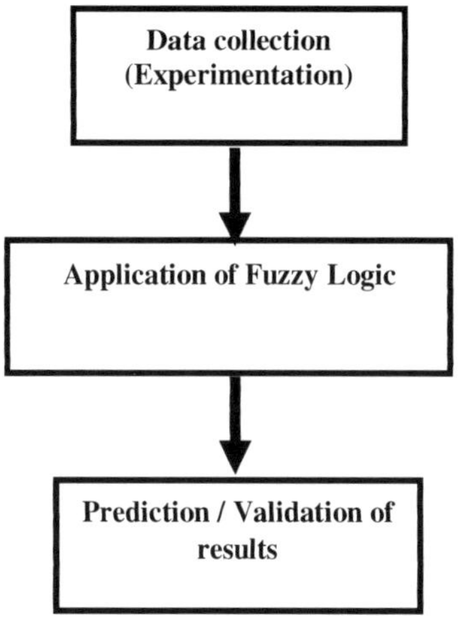

Fig.1. Scheme of Proposed methodology

## III. EXPERIMENTATION

Taguchi experimental design is employed to conduct experiments and to establish an correlation model between the input and output parameters and the optimization of response. The input factors are sliding velocity, applied load and distance travelled. Wear rate is the output variable. 27 experiments were performed under cutting environment by design of experiments. Wear experiments were carried out using computerized wear testing machine (pin on Disc) (DUCOM, TR-20LE-PHM-400) by ASTM-G99 standards. The design layout is as shown in Table 1. Before testing the pins and disc surface were cleaned with acetone. The wear test specimen (8 mm diameter and 20 mm length) was subjected to the load range 10,15,20 N at varying speed .

TABLE I. TAGUCHI DESIGN MATRIX

| Level | Sliding Velocity(m/s) | Distance Travelled(m) | Applied load (N) | Wear Rate (x10mm$^3$/m) |
|---|---|---|---|---|
| 1 | 1 | 200 | 10 | 30.45 |
| 2 | 1 | 200 | 10 | 30.5 |
| 3 | 1 | 200 | 10 | 31.62 |
| 4 | 1 | 350 | 15 | 35 |
| 5 | 1 | 350 | 15 | 34.09 |
| 6 | 1 | 350 | 15 | 33.69 |
| 7 | 1 | 500 | 20 | 41.02 |
| 8 | 1 | 500 | 20 | 39.06 |
| 9 | 1 | 500 | 20 | 40.05 |
| 10 | 1.5 | 200 | 15 | 35.02 |
| 11 | 1.5 | 200 | 15 | 35.05 |
| 12 | 1.5 | 200 | 15 | 35.07 |
| 13 | 1.5 | 350 | 20 | 40.06 |
| 14 | 1.5 | 350 | 20 | 40.79 |
| 15 | 1.5 | 350 | 20 | 40.33 |
| 16 | 1.5 | 500 | 10 | 30.12 |
| 17 | 1.5 | 500 | 10 | 30.32 |
| 18 | 1.5 | 500 | 10 | 30.33 |
| 19 | 2 | 200 | 20 | 41.03 |
| 20 | 2 | 200 | 20 | 40.95 |
| 21 | 2 | 200 | 20 | 41.11 |
| 22 | 2 | 350 | 10 | 31.09 |
| 23 | 2 | 350 | 10 | 31.11 |
| 24 | 2 | 350 | 10 | 31.32 |
| 25 | 2 | 500 | 15 | 36.09 |
| 26 | 2 | 500 | 15 | 36.14 |
| 27 | 2 | 500 | 15 | 35.99 |

## IV. DEVELOPMENT OF FUZZY LOGIC SYSTEM

The first step in fuzzy logic model is to accept the inputs and determine the degree to which they belong to each of the appropriate fuzzy sets through membership functions. In the proposed fuzzy logic model, the input is a crisp numerical value limited to the universe of discourse of the input parameter. The input factors are sliding velocity, applied load and distance travelled. The crisp values are listed in Table 2. The output function is a fuzzy degree of membership function is the qualifying linguistic set. Fuzzy logic modeling is based on rules and each of the rules leans on solving the inputs into a number of different fuzzy linguistic sets. Before the rules are set, the inputs are fuzzified to each of these linguistic sets. The inputs variables are fuzzified and the degree to which each part of the antecedent is accommodated for each rule. The input to the fuzzy operator is two or more membership values from fuzzified input factors. Membership functions for output factor is in Figure 2.Output variable is a single truth-value. Defuzzification method used in developing the model is centroid technique. Membership function for wear rate prediction, sliding velocity and distance travelled is shown in figures 3, 6 and 7. Rules and rule viewer is in figures 4 and 5.

TABLE II CRISP FUZZY VALUES

| Input | Crisp Values |
|---|---|
| Sliding velocity | 1 to 2 m/sec |
| Applied Load | 12 to 19 N |
| Distance travelled | 250 to 450 m |

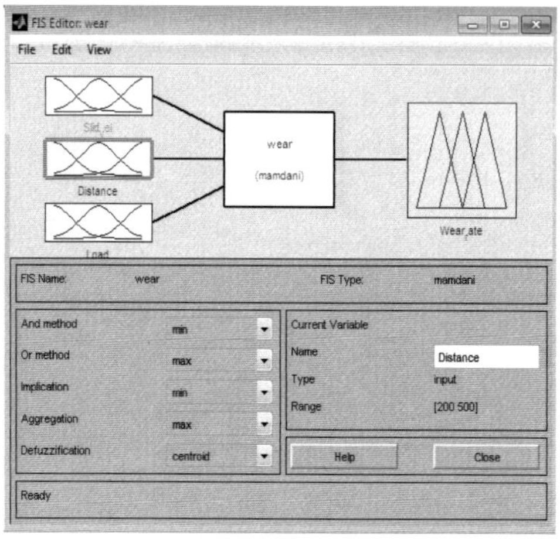


Fig.2. Developed Fuzzy logic structure

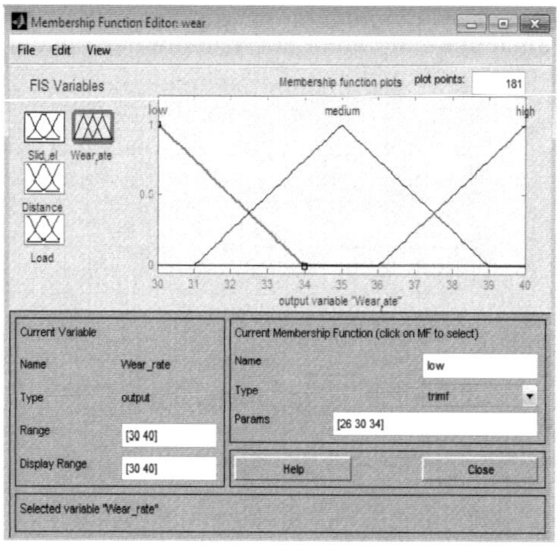


Fig.3. Membership for wear rate

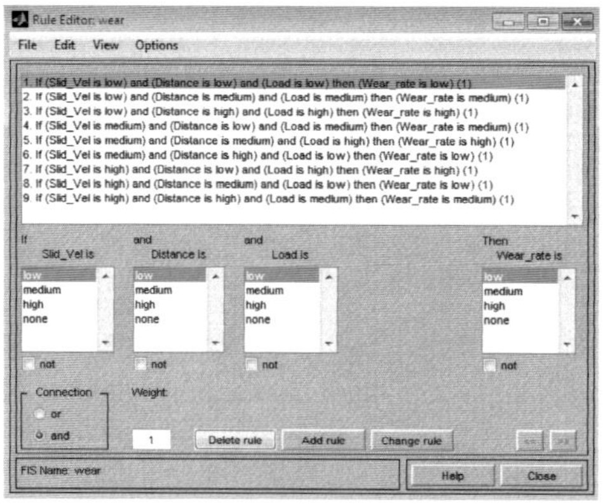


Fig.4. Rules for wear rate

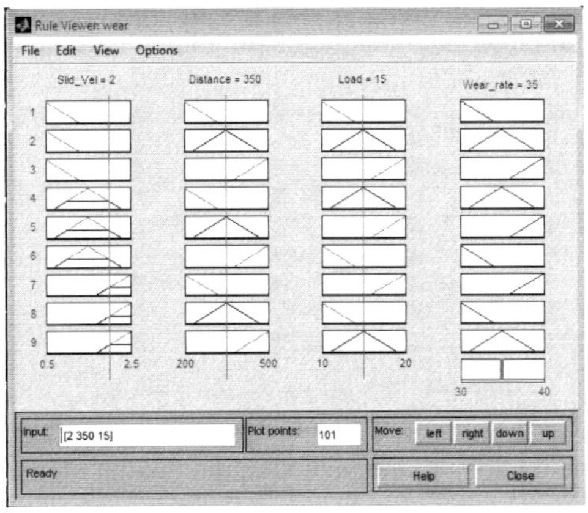


Fig. 5 Rule viewer for wear rate prediction

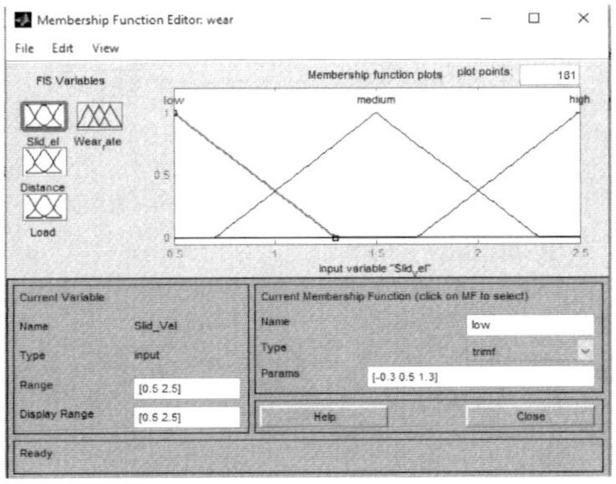


Fig.6. Membership for sliding velocity

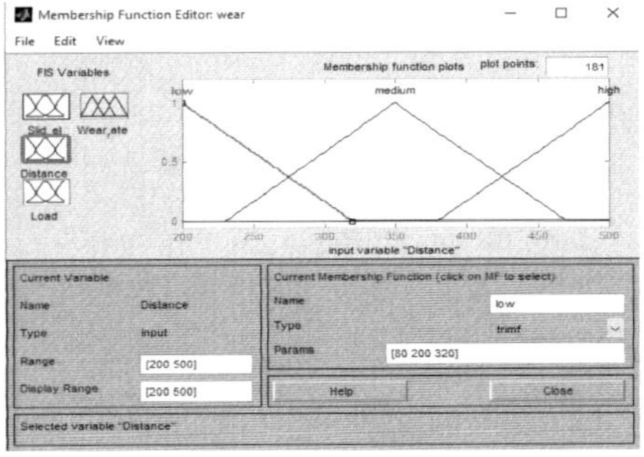


Fig.7. Membership for distance travelled

## V. Results and Discussion

The behaviour wear of Al 7075 titanium carbide composite fabricated by stir casting method was investigated using a pin-on-disc type wear rig by varying the parameters of sliding distance, applied load and distance travelled respectively. After developing the fuzzy logic system for the prediction of wear rate, the model is tested by giving various different values using fuzzy rule viewer.

If the input values are changed, the corresponding output is automatically obtained from the developed model. Confirmatory experiments are done to validate the developed fuzzy model and results are given in Table 3. Input parameters for the confirmatory experiments are given in Table 4. The errors in wear rate prediction very rarely exceed by 7 %, and thus the developed fuzzy model was able to predict with significant accuracy. The developed model is recommended and forwarded to predict wear rate under different conditions.

TABLE III CONFIRMATORY EXPERIMENT RESULTS

| Experimental values | Predicted values | % error |
|---|---|---|
| 32.06 | 34.13 | 6.065 |
| 37.16 | 34.91 | 6.445 |
| 41.07 | 39.77 | 3.268 |

TABLE IV INPUT VALUES

| Sliding velocity | Distance travelled | Axial load |
|---|---|---|
| 1 | 250 | 12 |
| 1.5 | 300 | 17 |
| 2 | 450 | 19 |

## VI. CONCLUSIONS

Experiments are designed and conducted develop fuzzy logic model to predict wear rate. The multiple performance dry sliding characteristics of aluminum hybrid composites was optimized using orthogonal array which is a procedure from Taguchi method. Observations from the proposed model are compared with experimental results. Results indicate that the developed model shows a good agreement with real results. Further confirmatory trials are carried out to validate the proposed approach. With these encouraging results the developed model can be further improved by including other cutting parameters such as vibration which affects wear rate. Electronic components are to be built up for online wear rate monitoring.

## REFERENCES

[1] Prashant Patil and E.R.Deore, "Prediction of Specific Wear rate of Glass and Bronze Filled PTFE Composites," International Journal of Current Engineering and Technology, 2015, Vol,5, No,6, 3901 – 3906.
[2] K.M. Subbaya , B. Suresha , N. Rajendra & Y.S. Varadarajan, "Taguchi approach for characterization of three-body abrasive wear of carbon-epoxy composite with and without SiC filler," Journal of Composite Interfaces, 2012, Pages 297-311.
[3] Deepak Bagle, Sanjay Shekhawat, Jitendra Chaudhari, "Wear Analysis of PTFE and its Composites under Dry Conditions using Design-Expert," International Journal of Scientific and Research Publications, Jan 2013, Volume 3, Issue 1.
[4] Nana Zhao, Yunhua Xu, Xing Huang, Lisheng Zhong, Jiuli Lu, "Microstructure and wear properties of niobium carbide particulates gradient-distribution composite layer fabricated in situ," Ceramics International, December 2016, Volume 42, Issue 16, Pages 18507-18515.
[5] Xu Liujie, Jiandong, Wei Shizhong, Zhang Yongzhen, Long Rui, "Optimization of chemical composition of high speed steel with high vanadium content for abrasive wear using an artificial neural network," Materials and Design, Dec 2005 , pp.1031-1037.
[6] Zhenyu Jiang, Zhong Zhang, Klaus Friedrich, "Prediction on wear properties of polymer composites with artificial neural networks", Composite science and technology," Institute for Composite Materials, Oct 2006, pp 168-176.
[7] B.V. Varade, Y.R. Kharde, "Prediction of Specific Wear rate of Glass Filled PTFE Composites by Artificial Neural Networks and Taguchi Approach," International Journal of Engineering Research and Applications, 2012, Vol. 2, Issue 6, pp.679-683.
[8] K. Sudarshan Rao, Y. S. Varadarajan and N. Rajendra, "Artificial Neural Network approach for the prediction of abrasive wear behavior of carbon fabric reinforced epoxy composite," 2014, Vol. 21, pp. 16-22.
[9] Zhenyu Jiang, "Neural network based prediction on mechanical and wear properties of short fibers reinforced polyamide composites," Material and Design, 2007, pp.628-637.
[10] JiahuaZhu, Yijun Shi, Xin Feng, Huaiyuan Wangi, Xiaohua Lu, "Prediction on tribological properties of carbon fiber and TiO2 synergistic reinforced polytetra-fluoroethylene composites with artificial neural networks," Materials and Design, 2008, pp.1042-1049.
[11] N. Leema, P. Radha, S.C. Vettivel, H. Khanna Nehemiah, "Characterization, pore size measurement and wear model of a sintered Cu–W nano composite using radial basis functional neural network," Materials & Design, 2015, Volume 68, Pages 195-206.

# Machining Chatter Analysis for High Speed Milling Operations

M. Sekar*
Department of Mechanical Engineering
GMR Institute of Technology, Rajam, India
mailtosekar@gmail.com

I. Kantharaj
Department of Mechanical Engineering
Karunya University, Coimbatore, India
kantharaj@karunya.edu

Savale Amit Siddhappa
Department of Mechanical Engineering
Karunya University, Coimbatore, India
amit.savale5@gmail.com

ABSTRACT

Chatter in high speed milling is characterized by time delay differential equations (DDE). Since closed form solution exists only for simple cases, the governing non-linear DDEs of chatter problems are solved by various numerical methods. Custom codes to solve DDEs are tedious to build, implement and not error free and robust. On the other hand, software packages provide solution to DDEs, however they are not straight forward to implement. In this paper an easy way to solve DDE of chatter in milling is proposed and implemented with MATLAB. Time domain solution permits the study and model of non-linear effects of chatter vibration with ease. Time domain results are presented for various stable and unstable conditions of cut and compared with stability lobe diagrams.

Keywords- Delay Differential equations, Regenerative Chatter, R-K method.

## I. INTRODUCTION

The term High Speed Machining (HSM) commonly refers to end milling at high rotational speeds and high surface feeds [1]. High speed machining (HSM) is increasingly used to machine complex profiles with a goal of reducing lead times [2]. Machining Aero Engine parts and thin walled Jet Engine components in HSM is always prone to non-linear chatter vibrations as geometry changes abruptly and acceleration becomes out of machine tool limits. The ever increasing need of productivity and quality has made chatter an important area of research since 1960s. Machine tool chatter arises due to forced and self-excited vibrations. Tlusty [1] proved that the regenerative effect due to feedback effects of work-piece geometry from the previous tool pass is the major source of machine tool chatter. Regenerative chatter occurs when the cutting tool removes a surface with undulations created by an earlier tool pass. The associated change in chip thickness results in a varying dynamic cutting force resulting in a wavy surface accompanied with vibration which eventually leads to unstable machining process. Merrit [2] modeled this process in the form of a block diagram with feedback loop and solved using linear control theory. The resulting stability conditions are expressed in the form of stability lobe diagrams proposed by Tobais [3] which depict the maximum chatter-free chip widths as a function of spindle speeds. The equation to determine the stability conditions for turning operations is simple as the direction of forces is time independent. However for milling the rotating cutter causes the direction of chip generation to vary continuously resulting in time dependent coefficients. Following the work of Sridhar et al. [4] and Minis

* Corresponding author

et al. [5], Altintas and Budak [6] proposed a feasible formula to estimate the stability condition by expressing the coefficients with a truncated Fourier series.

Though the above approach is widely used, in recent times to obtain deeper insight to chatter other than the chatter stability conditions, time domain solutions of equations of motion within the plane of cut are studied using numerical integration solved in time domain [7-8]. With the advances in the modern theory on non-linear dynamics and latest computer hardware it is possible to analyze the above problems easily. In addition, time domain simulation methods are also preferred to study low immersion cuts where a combined analysis of flexible tool and work piece system is required [9]. The time domain analysis helps to control vibration actively [10]. The mathematical model for chatter vibration is best described by time delay differential equations characterized as two parts: 1) the ordinary differential equation describing periodic motion and chatter free condition and 2) the delay differential part describing the chatter area [9]. Since closed form solutions of delay differential equations are difficult, numerical solutions are considered as the most important techniques in time domain problems involving such dynamics [11]. Noticeable improvements are adaptive integration, estimation and control of local through variable step size [12]. Since a continuous solution rather than discrete is required, interpolants also play an important role in solving the delay differential equation [13]. Li et al. [7] implemented R-K method to compute the time domain solution for chatter problems. In this approach the step size remained constant and error control is not implemented which are not advisable according to modern R-K literatures. Though modern DDE solvers with adaptive step size R-K and error control algorithms exists, as far as per authors' knowledge they are not implemented for chatter problems.

The objective of the paper is to adapt a well-developed DDE code to solve chatter problems. The DDE23 solver of MATLAB is chosen to solve the chatter equations. DDE23 was developed by Shampine [14, 15] to suit a wide range of delay differential problems. The customization of DDE solver to model chatter problems is explained in sections 4 & 5. Numerical simulation results are presented in Section 6 and conclusions are given in section 7.

## II. CHATTER VIBRATION EQUATIONS

In this section the mechanics of end-milling process and the analytical model of a two-degree of freedom flexible tool system are presented. This model considers self-excited regenerative mechanism as the principle source of chatter. The regenerative chatter is a phenomenon associated with the modulation of chip thickness due to feedback between subsequent tool passes. Modulation of chip thickness results in oscillatory cutting force which causes a periodic oscillatory motion of the work-piece-tool system and can be characterized by different modes of vibration. It is generally considered, modes perpendicular along the cutter axis are dominant and hence chatter in milling is expressed as 2 DOF delay differential equation.

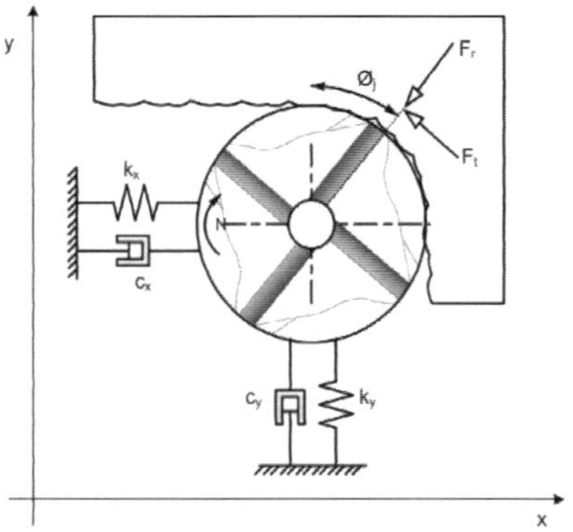


Fig. 1. Two DOF flexible end milling cutter model.

[Type here]

*A. Equations of motion*

The flexible tool model is assumed to be a spring mass system with dampers with 2 degrees of freedom. For simplicity the work-piece is considered to be rigid. Fig. 1 shows the 2 DOF-model of flexible multi-teeth milling cutter with simultaneous engagement. Simultaneous engagement refers to engagement of more than one cutting edge of the milling cutter. The tool is assumed to have straight flutes for simplicity in modeling of cutting forces. The radial and tangential cutting forces ($F_r$, $F_t$) acting on a tooth can be expressed using the cutting coefficients ($K_t$, Kr) and chip width (h). The material dependent cutting coefficients can be found from standard machining data handbooks [1].

In local coordinates the forces ($F_r$, $F_t$) are

$$\begin{Bmatrix} F_t \\ F_r \end{Bmatrix}_j = \begin{bmatrix} K_t\, h(\theta)\, a \\ K_t\, k_r\, h(\theta)\, a \end{bmatrix}_j = \begin{bmatrix} K_t\, h(\theta)\, a \\ F_t\, k_r \end{bmatrix}_j \qquad (1)$$

where θ is the instantaneous angle made by the rotating end mill tooth (z-axis) referenced with respect to y axis and 'a' is the axial depth of cut (ADOC).

Transformation of Eqn. 1 to global coordinates results in resolution of forces on X and Y direction as shown below:

$$\begin{Bmatrix} F_x \\ F_y \end{Bmatrix}_j = \begin{bmatrix} -sin(\theta) & -cos(\theta) \\ -cos(\theta) & sin(\theta) \end{bmatrix}_j \begin{Bmatrix} F_r \\ F_t \end{Bmatrix}_j \qquad (2)$$

The cutting tool dynamics of the end mill cutter is influenced by the parameters namely mass, stiffness and damping factor. The periodical vibrating force is induced by the regenerative effect of the tool. The equations of motion of the flexible tool system consisting of Z cutting edges can be written as

$$\ddot{x} + 2\zeta_x \omega_{nx} \dot{x} + \omega_{nx}^2 x = \frac{1}{m_x} \sum_{j=0}^{Z-1} F_{xj} = \frac{1}{m_x} F_x(t) \qquad (3)$$

$$\ddot{y} + 2\zeta_y \omega_{ny} \dot{y} + \omega_{ny}^2 y = \frac{1}{m_y} \sum_{j=0}^{Z-1} F_{xj} = \frac{1}{m_y} F_y(t) \qquad (4)$$

where is $m$ the modal mass, ξ is the damping ratio and $\omega_n$ is the natural angular frequency and F is the cutting force. Depending on the immersion angle of the cut, a Z toothed cutter may engage with multiple teeth on the work piece. Hence the forces on every tooth computed are summed up to consider the effect of all the teeth engaged in the cut as shown in Eqns. 3 & 4.

*B. Computation of Chip thickness*

Evaluation of instantaneous forces requires computation of chip thickness. The instantaneous chip thickness h(θ) due to regenerative effect is calculated as follows.

In case of a smooth surface the chip thickness can be expressed as

$$h_s = c\, sin(\theta) \qquad (5)$$

where c is the chip load or feed per tooth. Since the chip thickness changes due to the feedback effect of the previous pass (t-T), it is expressed in local coordinates u-v (u: radial direction) as

$$h_d = h_s + u(t-T) - u(t) \qquad (6)$$

where T is the tooth passing period given by

[Type here]

IOP Conf. Series: Materials Science and Engineering **247** (2017) 012014 doi:10.1088/1757-899X/247/1/012014

$$T = \frac{60}{NZ} \quad (7)$$

with N as the cutter speed in minutes. Though it is possible to use the $h_d$ directly to compute forces, for convenience it is expressed in the x-y global coordinate system using the transformation matrix.

$$u(t) = -x(t)\sin\theta - y(t)\cos\theta \quad (8)$$

$$u(t-T) = -x(t-T)\sin\theta - y(t-T)\cos\theta \quad (9)$$

Substituting in the Eqn. 6 for $h_d$

$$h_d = (h_s + [x(t) - x(t-T)]\sin(\theta_i) + [y(t) - y(t-T)]\cos(\theta_i)) \quad (10)$$

The above equation shows that the chip thickness is dependent of previous passes which makes the equation into a Delay Differential Equation (DDE).

*C. Effect of intermittent cutting*

In machining with end-mills, the tool edge does not engage with work all the times. The tool engagement time is dependent on the angle of entry and exit. This produces intermittent cutting operation. To consider the effect of intermittent cutting action of the tooth of the milling cutter a unit step function $g(\theta_i)$ is multiplied with the above equation. It must be noted that at any given point of time, depending on the number of tool edges and entry and exit angle, more than one cutting edge may be involved in the cutting operation.

$$g(\theta) = \begin{cases} 1 & \theta_{entry} < \theta < \theta_{exit} \\ 0 & otherwise \end{cases} \quad (11)$$

The angular position at any given instant t can be found by using the relation

$$\theta(t) = \frac{2\pi Nt}{60} + \frac{Z2\pi}{N} \quad (12)$$

Use Eqn. 12 the instantaneous position of the tool edge can be computed. If it lies between the entry and exit angle of the work, forces can be obtained according to Eqn. 2 otherwise forces experienced from that edge will be zero according to Eqn. 11. Substituting the $h_d$ in the Eqn. 1 forces $F_x$ and $F_y$ can be computed from eqn.2. Subsequently eqns. 3 & 4 can be solved numerically using R-K method.

## III. STATE SPACE FORM OF DDE

To solve a higher order differential equation, it must first be expressed as first order simultaneous differential equations. A state space form which is easy to implement in MATLAB software is used here. For the chatter equations (Eqns. 3 & 4) the state variables can be expressed as follows

$$x_1 = x \quad (13)$$

$$x_2 = \dot{x}_1 \quad (14)$$

and the simultaneous differential equations can be written as

$$\dot{x}_1 = x_2 \text{ and} \quad (15)$$

$$\dot{x}_2 = \frac{1}{m}(F - cx_2 - kx_1) \quad (16)$$

[Type here]

IOP Conf. Series: Materials Science and Engineering **247** (2017) 012014 doi:10.1088/1757-899X/247/1/012014

## IV. DELAY DIFFERENTIAL EQUATIONS

Delay differential arise in all types of disciplines and it has received wide attention which has lead to the development of efficient solvers. Though DDEs are similar to ODEs, their solution procedure is different as the solution is dependent on its previous delay period(s). The delay period in chatter problems is a positive real number and multiple delay periods are common in chatter DDEs. Once the solution for the current delay period is computed, the sequence is continued for the next delay period. The solution at the previous time period is called as history in DDE terminology. While solving DDEs, discontinuous derivates may arise and require special treatment. Since the effect of low order discontinuities is more, it is tracked till it is smoothened out. Solvers usually calculate the discontinuities in advance and truncate the higher order discontinuities [15].

Runge-Kutta method is the most widely method used to solve DDEs however when used with constant step size it leads to poor error control. Modern solvers use variable step size, though choosing a proper step size without increasing computation burden is an important issue. Moreover it must be chosen to get an accurate approximation. Since DDE with variable step size require solution at previous times other than mesh points, interpolation of results is required. Cubic hermite interpolants are widely used in DDE solvers to obtain the solution everywhere in the interval. In this paper the solution for DDE is obtained using DDE23 which is proposed by Shampine [14]. The solver uses BS (23) pair with continuous extensions of R-K formulas which enables to compute solution at any given time 't'. DDE23 also provides an powerful event capability which is useful to solve milling chatter problems with multiple cutting edges simultaneously acting together. However DDE23 is not suitable for variable time delays problems that occur due to varying speed and varying pitch.

## V. SOLUTION PROCEDURE

### A. DDE interface

In this section the implementation of chatter equations for DDE23 solver is explained using its standard interface. The dde23 function call is as follows

*sol=dde23('ChtrddeFunc', [time delay], 'ChtrHistory', [tintial, tfinal], options, state);* (17)

where

1) *ChtrddeFunc* : a handle to function that defines the state space form of a delay differential equation (Eqns. 19 and 20).
2) *[Time delay]:* time delay which is equal to T according to Eqn. 7.
3) *ChtrHistory*: a handle to function for solution history
4) *[Tintial tfinal]:* the time span of integration which is user defined. Generally 0-3 seconds
5) *Options:* used for reentry
6) *State:* variable passed to trigger the start of a new integration.

The dde23 solver returns the solution '*sol*' which is a structure consisting of time, solution 'y' and its derivative. The dde23 solver updates time 't' automatically and the adaptive increment to parameter 't' is chosen by the dde23.

### B. Modeling chatter as function

The first step to solve chatter equations consists of implementing the equations mentioned in section 2 as a function *'chtrddefunction'*. The function comprises the chatter equations (Eqns. 17-20) expressed in state space form. To compute the forces at a given instant, 't' the angular position of the tooth (θ) has to be first found. Employing Eqn.12 and looping through 1 to Z number of teeth, the instantaneous positions of the all the teeth can be calculated. Though the initial position can be at any random angle, for simplicity the initial position is assumed to coincide with positive Y axis. When the location angle 'θ' lies between the '$\theta_{entry}$' and '$\theta_{exit}$' forces $F_x$, $F_y$ are computed otherwise they are set to zero according to Eqn. 11. This sub- procedure is made as a loop and all the forces are summed according to the Eqns.3 & 4.

[Type here]

The chip thickness is computed according to Eqn.10. The history of previous values at (t-T) is provided by the Z lag values which are passed to this function as an input parameter according to the syntax.

Table .1: Data for MATLAB simulation

| | |
|---|---|
| Natural freq. $\omega_{nx}$, $\omega_{ny}$ (Hz) | 600, 660 |
| Damping ratio $\xi_x$ ,$\xi_y$ | 0.035, 0.035 |
| Stiffness $k_x$ $k_y$ (N/m) | $5600*10^3$ |
| Number of teeth | 4 |
| $K_t$ (MPa), $K_r$ | 600, 0.07 |
| Milling type | Half immersion |

*C. Defining history of chatter*

The second step is to define the history function. In most of the chatter problems the initial displacement profile is assumed to be '0'. With this approach, it is possible to express the initial profile and its derivates using the history function.

*v = ChtrHistory (t, state)* (18)

where 'v' is the vector of state variables computed for time 't' . The history definition is used only during the first iteration and for subsequent integration histories; the values from DDE solution (Z lag values) are used.

*D. Restarting integration after each tooth pass*

The next step is concerned with restarting of the integration. The integration of DDE cannot be continued indefinitely after every tooth passing period since the newly engaged tooth will encounter the profile left by the previous tooth. Hence the solution for the current and newly engaged tooth must consider the displacement profile left by the previous tooth and its derivative. This can be modeled using the event options of DDE23 through the ddeset command.

*options = ddeset('Event','ChtrEvents');* (19)

where '*ChtrEvents*' is a handle to the function which is similar to '*Chtrddefunction*' function. The function compares the state value with the instantaneous time 't' and when they become equal, an event is triggered to stop the current integration.

To obtain a meaningful result, the numerical integration process must be executed for a certain time span set by the user. During this time, the dde23 solver is called repetitively in a while loop at time intervals equal to the tooth passing period. In each iteration, the parameter 'state' is incremented in integer increments of time delay period. Thus the events function triggers the integration to stop when the state values equals the tooth passing period and the solution is placed in *sol* variable. Comparing the time 't' from the solution structure with the given max time span the execution of while loop can be controlled. For all the subsequent calls made by dde23 solver, the solution returned by the previous call is passed as history. The starting value of the time span is updated with the final time value available at the end of solution structure. The summary is shown in figure 2.

[Type here]

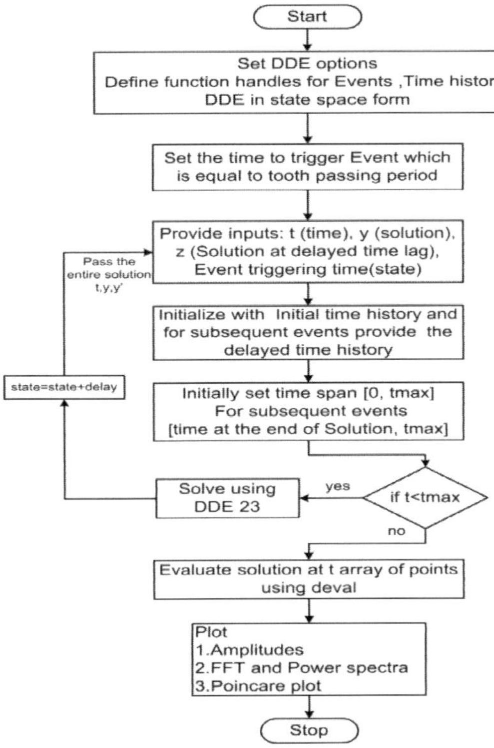


Fig. 2. Summary of the procedure

## VI .RESULTS AND DISCUSSION

Table 1 shows the data used to evaluate the proposed procedure. A straight fluted end milling cutter is used for simplicity. Fig. 3 shows the trimmed stability lobe diagram computed using the conventional approach with these parameters [16, 17]. The diagram can be divided into two regions namely 1) stable region with no chatter and 2) unstable regions with chatter. It must be noted that non-linearity effects are ignored in this approach to arrive closed m solution of onset of chatter. Using procedures mentioned in the previous section, the chatter delay differential equations are solved and the time domain solutions are computed for various speeds and depth of cut. Figs. 4 & 5 show the displacement profile along X and Y direction computed for a speed of 3000 rpm with depth of cut 1mm. From these figures one can find that the diverging profile leads to instability resulting in chatter.

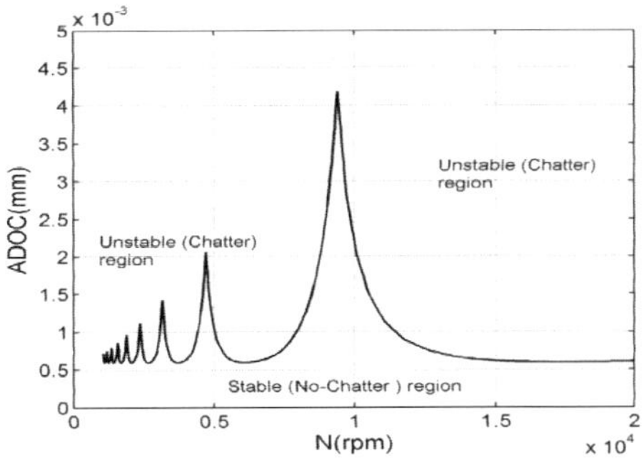


Fig. 3. Stability lobe diagram using Table 1

[Type here]

IOP Conf. Series: Materials Science and Engineering **247** (2017) 012014 doi:10.1088/1757-899X/247/1/012014

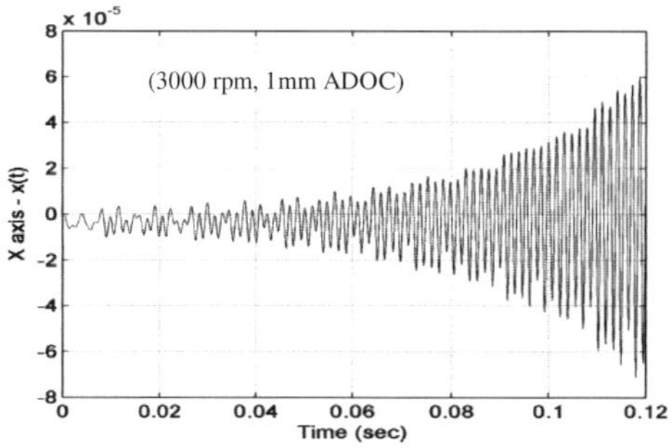


Fig. 4. X-Displacement profile

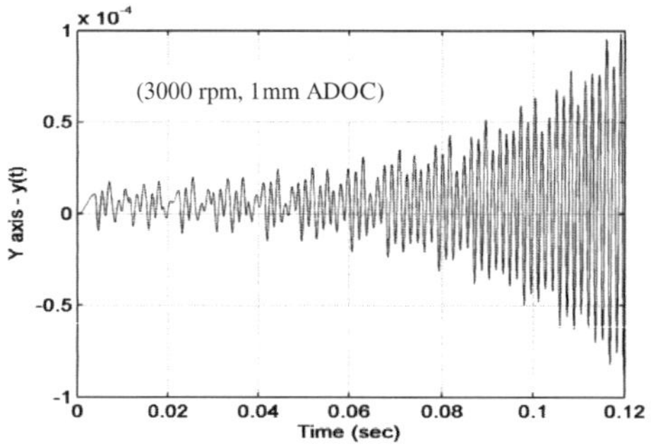


Fig. 5. Y-Displacement profile

However for the speed of 10000 rpm with 3mm depth of cut, the X displacement profile clearly shows diverging characteristics though the amplitude is very small (Fig. 6). The Y-axis profile shows a near stable profile though its diverging trends are not visible (Fig. 7). However, according to conventional stability lobe diagram (Fig. 3), this data point lies in the stable region. This shows the difference between the time domain approach and stability lobe approach since the later does not model the non-linearities. A stable cut at 15000 rpm and 0.5 mm depth of cut is observed (Figure 8). The computation time for all the above cases is observed to average between 30 seconds and 40 seconds with a Pentium Celeron processor with 2GB RAM. Due to variable step size in solving the chatter delay differential equation, the time taken to obtain time domain solution varies from case to case. This method is preferred for constant delays problems which is a common in high speed machining.

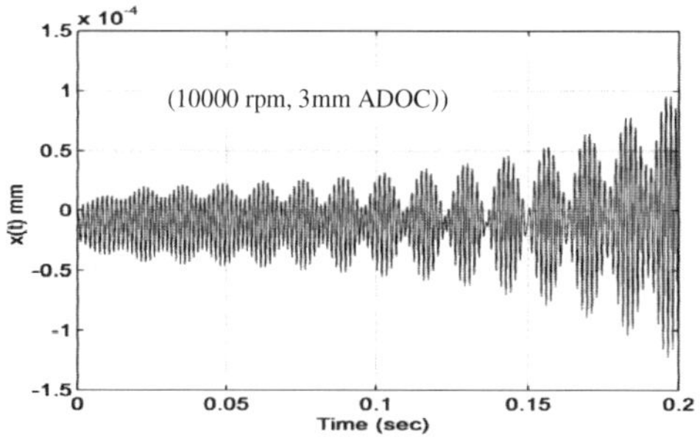


Fig. 6. Displacement profile (10000 rpm, 3mm ADOC)

(10000 rpm, 3mm ADOC))

[Type here]

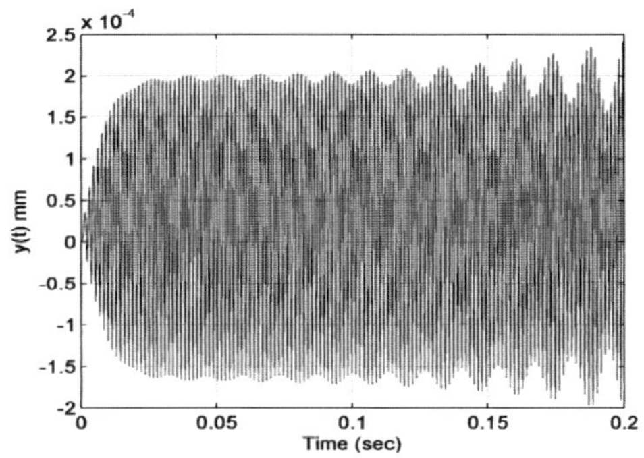


Fig. 7. Y-Displacement profile

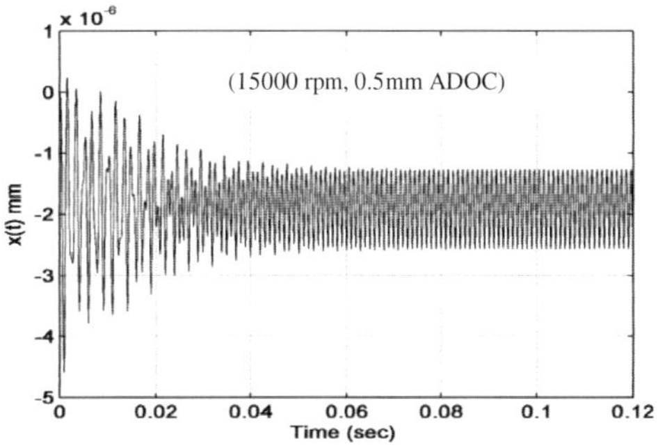


Fig. 8. X-Displacement profile

Further, it is possible to analyze the chatter conditions according to frequency spectrum computed from the displacement profile. From the Fig. 10 it can be noticed that the multiple frequency peaks leads to chatter condition. A clear peak frequency corresponding to the tool path frequency indicates chatter free machining process. With simple modification the proposed method can be extended to analyze chatter in turning process.

The procedures developed in this article will be helpful for the study of non-linear chatter problems in high speed machining. The tool displacement profiles along x and y provide inputs for constructing bifurcation diagram and the relevance of onset of chatter can be analyzed based on floquet theory. Useful inferences can be made by comparing the results with conventional stability lobe diagram. In addition the time history of the previous periods provide useful information for analysis.

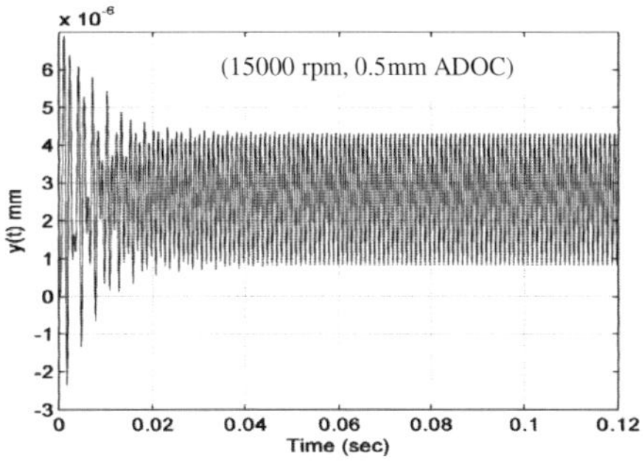


Fig. 9. Y-Displacement profile

IOP Conf. Series: Materials Science and Engineering **247** (2017) 012014 doi:10.1088/1757-899X/247/1/012014

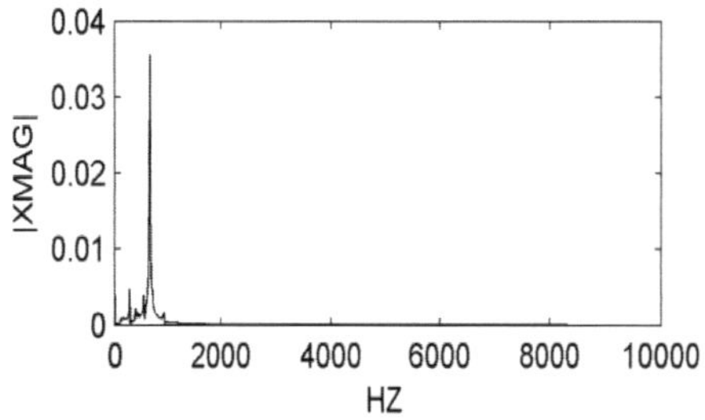


(a) Chatter FFT profile showing multiple frequencies (3000 rpm, 1mm ADOC)

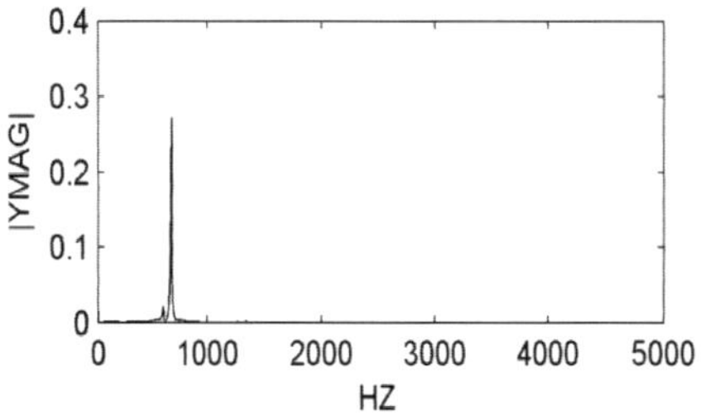


(c) No chatter FFT profile showing one dominant frequency (10000 rpm, 3mm ADOC)

Fig. 10. Analysis of chatter from FFT profiles from time domain results

## VII. CONCLUSIONS

1. A new approach to obtain time domain solution to chatter problems with non-linearities is proposed in this paper. A time domain solution for an example problem is illustrated.
2. The solver used in this method equips with adaptive time step size and error control methods resulting in a more accurate solution. It is suitable for multiple time delays and problems with added DOFs. It is possible to construct Poincare plot as derivatives are available in the solution structure.
3. Though a straight flute cutter is used for analysis, the proposed method can be extended to helical flute cutter by modifying the program segment that computes forces.
4. The proposed method is suitable to solve problem with an initial profile.
5. The proposed method can be used as an alternate to conventional stability lobe diagram with loops.

## REFERENCES

[1] H. Schulz and T. Moriwaki, "High-speed Machining", Annals of the ClRP, Vol. 41,pp.637-643,1992

[2] M. Sekar, V.N. Narayanan and Seung-Han Yang, "Design of jerk bounded feedrate with ripple effect for adaptive NURBS interpolator", International Journal of Advanced Manufacturing Technology,Vol.37, No. 5 pp. 545–552, 2008.

[3] J. Tlusty, "Manufacturing Process and Equipment," Prentice Hall, USA, 2000.

[4] H.E. Meritt, "Theory of Self-Excited Machine Tool Chatter," ASME J. Engg. Industry, 87, pp. 447-454, 1965.

[5] S.A. Tobias, "Machine tool Vibration," Blackie, London, 1965

[6] R. Sridhar, R.E. Hohn, and W. Long, "A General Formulation of the Milling Process Equation," ASME J. Engg Industry, 90, pp. 317-324, 1968.

[Type here]

[7] I. Minis and R. Yanushevsky, "A new theoretical approach for the prediction of machine tool chatter in milling," ASME J.Engg Industry, 115 , pp. 1-8, 1993.

[8] Y. Altintas and E. Budak, "Analytical prediction of stability lobes in milling," Annals of CIRP, Vol. 44, pp. 357-362, 1995.

[9] M.X. Zhao and B. Balachandran, "Dynamics and stability of milling process," Int. J of solids and structures, Vol. 38, pp. 2233-2248, 2001.

[10] H. Li., and X. Li, "Modeling of simulation of chatter in milling using a predictive force model,", Int .J. Mach Tools and Manuf., Vol. 40, pp. 2047-2071, 2000.

[11] T. Insperger, J. Gradisek, M. Kalveram, G. Stepan, K. Winert and E. Govekar, "Machine Tool Chatter and Surface Location Error in Milling Processes," ASME Journal of Mfg Science, Vol.128, pp. 913-920, 2006.

[12] W. Andrzej, R. Rafal and W. Jerzy, "The concept of active elimination of vibrations in milling process", Procedia CIRP, Vol. 31, pp. 82 – 87, 2015.

[13] J. H. E. Cartwright and O. Prio, "Dynamics of Runge-Kutta methods," Int. Journal of Bifurcation and Chaos, Vol. 2, pp. 427-449, 2002

[14] W. H. Enright, D.J. Higham, B. Owren and P.W. Sharp, "A survey of Explicit R-K method," TR 291/94, Depart. of Comp. Sci., University of Toronto, 1994.

[15] C.T.H. Baker, C.A.H. Paul and D.R. Willé, "Issues in the Numerical Solution of Evolutionary Delay Differential Equations," Adv. Comput. Math., 3, pp. 171–196,1995.

[16] L.F. Shampine and S. Thomson"Solving DDEs in MATLAB," Applied Numerical Mathematics, 37, pp. 441–458, 2001.

[17] L.F. Shampine, I. Gladwell, S. Thomson, "Solving ODES with MATLAB," Cambridge University press, UK, 2003.

[18] M. Sekar, J. Srinivas, N. Kang and S.H. Yang, "An Integrated Approach to the Improvement of Stability Lobes," Int.. J. of Precision Engineering and Manufacturing, Vol. 9, pp: 83-85, 2008.

[19] M. Sekar, J. Srinivas, K.R. Rama Kotaiah and Seung-Han Yang, "Stability Analysis of Turning Process with Tailstock-Supported Workpiece," Int. J. of Advanced manufacturing Technology, Vol. 43, pp. 862-871, 2009.

[Type here]

IOP Conf. Series: Materials Science and Engineering **247** (2017) 012015 doi:10.1088/1757-899X/247/1/012015

# Influence of tool pin in friction stir welding on activated carbon reinforced aluminium metal matrix composite

G.DijuSamuel[a],J.Edwin Raja Dhas[b]

[a]Research Scholar , Department of Aeronautical Engineering, Noorul Islam University, Kumaracoil,629180,India

[b]Professor and Head, Department of Automobile Engineering,Noorul IslamUniversity,Kumaracoil,629180,India

Corresponding author: dijusams@gmail.com , +919994441656

**Abstract**: This paper focus on impact of tool pin in friction stir welding on activated carbon reinforced aluminium metal matrix composite. For fabrication of metal matrix composite AA6061 is used as matrix and activated carbon is used as reinforcement and it is casted using modified stir casting technique. After casting metal matrix composite has undergone various microstructure tests like SEM,EDAX and XRD. FSW is carried out in this metal matrix composite by choosing various tool pin profile like square,round,Threaded round, hexagon and taper. The quality of welded plates is measured in terms of ultimate tensile strength and hardness.

*Keywords: AA6061, SEM, EDAX, XRD, FSW, Hardness, Tensile Strength.*

## 1.INTRODUCTION

Metal matrix composites (MMCs) are being an unavoidable metal in aviation field because of the benefit of low weight to quality proportion.MMCs are one of the most important class of composites for structural, thermal and kinetic applications. Aluminium based hybrid composites have good strength, low weight, good weldability and corrosion resistance. It is widely used in industries like aerospace,automobile,marine etc [1]&[2].

Stir casting procedure is a fluid state technique for the creation of metal matrix composites in which clay particles (reinforcement) are blended with a liquid metal(matrix) by methods for mechanical mixing [3-5].In that observation aluminium fly ash composite have improved their mechanical properties by adding some suitable percentage of reinforcement.[6].In some other literature there will be changes in mechanical properties by mixing reinforcements like SiC,Ti,Albite, graphite etc.Tensile and compressive property was improved by adding AA6061 with different weight percentage of albite and graphite[7].

AMC can be used in this welding since in this process rotating tool moves alongjoint interface, produces warmth and brings about recycling stream of plasticized material close to the tool surface. This mellowed material is subjected to expulsion by the tool pin rotational and navigate developments prompting arrangement of friction stir processing zone [8].

During this welding, a special shaped FSW tool is rotated, plunged and then traversed along the joint to form weld [9] FSW has been used to weld the monolithic materials such as aluminum, magnesium, steel. Recently, several studies have reported that the FSW produces a high quality weld jointin aluminum based composites [10].The advantage of friction stir welding is it will eliminate all types fusion problem during welding.From literature [11-15] it is understood there will be influence in tool pin profileused in friction stir welding in literature pin profile like square, hexagon, taper, round, threaded round, Threaded Taper,Triangle are used from this four influential profile is used in this research work. In this paper the welded composite plates mechanical properties are measured.

## 2. Experimentation:

### 2.1 Friction Stir Casting :

The materials used to fabricate metal matrix composite are AA6061 rod and powdered activated carbon.The matrix for the composite is chosen as AA6061 cylindrical rods and activated carbon is used as reinforcement. Aluminum bars were put in a graphite cauldron and were covered to dodge pollution and warmed utilizing an electrical furnance. The AA6061 rods were melted at a temperature of 900°C utilizing an electric heater.The liquid compound is mixed utilizing mechanical stirrer to frame a fine vortex.The Preheated activated carbon powder at a temperature of 600°C for a hour were then included at a consistent bolster rate into the liquid aluminum. 1 wt. % of magnesium particles was used as wetting agent. The composite is prepared by adding 6% of reinforcement, 93% of matrix and 1% wetting agent. From literature it is evident there will be no proper mixing of reinforcement beyond 6% of activated carbon. The composite obtained after casting is shown in figure.1

Figure.1 AA6061/activated carbon metal matrix composite

2.2 Microstructure test:

The fabricated metal matrix composites will undergo microstructure test to confirm the proper mixing of matrix and reinforcement. The presence of reinforcement also confirmed by EDAX and XRD tests.

IOP Conf. Series: Materials Science and Engineering **247** (2017) 012015 doi:10.1088/1757-899X/247/1/012015

## 2.3 SEM TEST

The Fig 2 a represents the microstructure of AA 6061/activated carbon metal matrix composite.The figures obviously uncovers the homogeneous circulation of the enacted carbon in the Al compound matrix and there is no confirmation of porosity and splits in the castings. The presence of matrix and reinforcement is highlighted in the figure.2

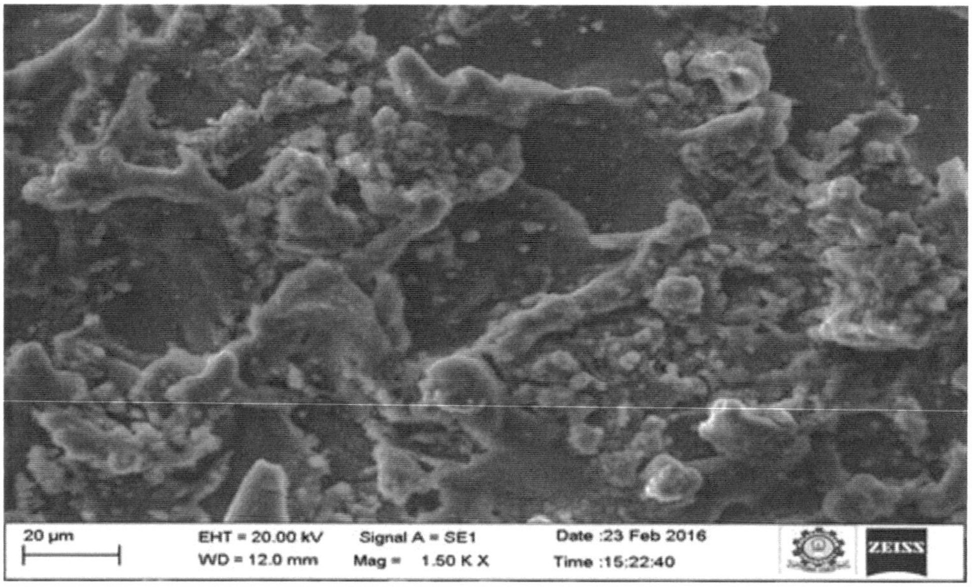


Figure.2 SEM image AA6061/activated carbon metal matrix composite

## 2.4 EDAX TEST

From fig 3EDS spectrum shows both the occurrence of Al alloy and activated carbon in the fabricated compound. The micro graph also shows that there is no agglomeration of the aluminum and activated carbon particles in the mixture.It is seen from this, that the exchange and mechanical blending of materials have occurred between two sliding surfaces, prompting the arrangement of a mechanically blended layer on the ragged surfaces.

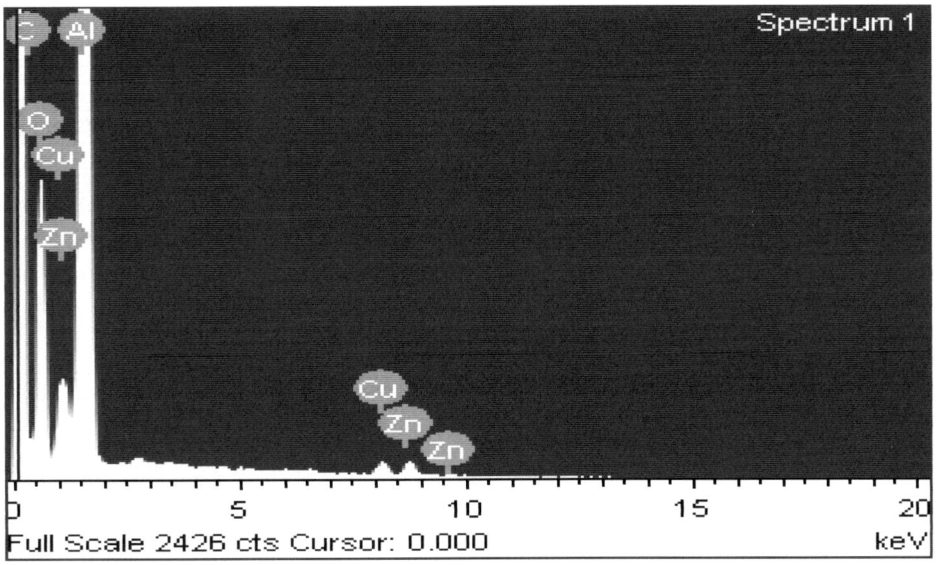


Figure.3 EDAX of AA6061/activated carbon metal matrix composite

## 2.5 XRD TEST

X- Ray Diffraction (XRD) of fabricated AA6061 and activated carbon metal matrix composite is shown in figure 4. It is observed that, peaks corresponding to aluminium and activated are present in all the three materials, whereas activated carbon peaks were present in AA6061/ activated carbon composites and no other reaction product peaks are seen. xrd results shows there is no formation other compound and also shows that it has good mixture**.**The resultspecifies the occurrence of aluminum in the major peak and the activated carbon in the second largest peak.In this an obviously noticeable activated carbon pinnacle can be seen in the AA6061 composite.The expansion in the force of the activated carbon tops with the aluminum substance of the composite is apparent.

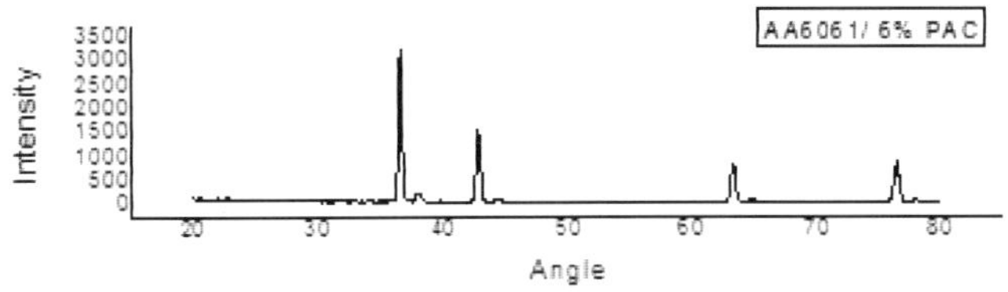


Figure.4- X-Ray Diffraction results of casted AA6061/PAC Composite

## 3.Friction Stir Welding:

The plate of dimensions 100mm×50×6mm are obtained in modified stir casting with various weight rate of reinforcement and different matrix materials are friction stir welded in modified vertical milling machine. The butt welding is carried out in AA6061 reinforced activated carbon composite plates.Tool selected for performing welding is High speed steel tool with four different probe shapes such as Square,Taper, Triangle, Round, hexagon and threaded round with tool length 70mm, shoulder diameter 16mm and probe length of

IOP Conf. Series: Materials Science and Engineering **247** (2017) 012015 doi:10.1088/1757-899X/247/1/012015

5.8mm.From literature the parameters for friction stir welding are spindle speed, axial load , tilt angle and tool profile.Among this four parameters tool profile will be varied other three parameters are kept constant.The values chosen are spindle speed 1200rpm,tilt angle zero and axial load 10KN. The process parameter for friction stir welding is shown in table.1.The specimen after welding were taken to tensile and hardness test as per ASTM standards.The figure 5 shows the five different types tools used for welding. Figure 6 shows friction stir welded composite plates by using different tool profile.

Table.1 Process parameters for friction stir welding

| Sl.No | Parameter | Range |
|---|---|---|
| 1. | Spindle speed | 1200rpm |
| 2. | Axial load | 10KN |
| 3. | Tilt angle | zero |
| 4. | Tool pin profile | Round, T. Round, Taper, Square Hexagon |

**Figure 5 Five different tools for friction stir welding**

IOP Conf. Series: Materials Science and Engineering **247** (2017) 012015 doi:10.1088/1757-899X/247/1/012015

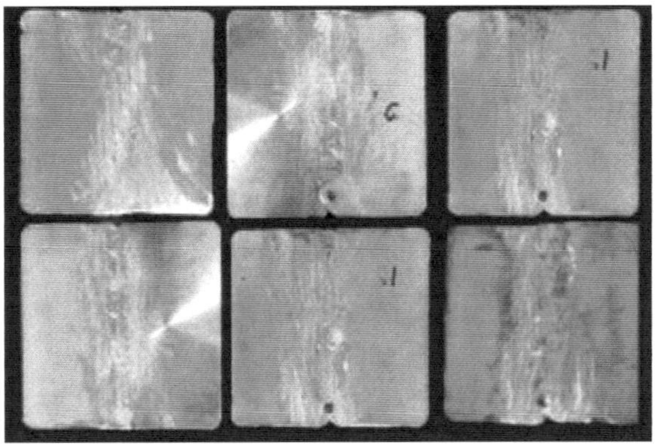

**Figure 6 Friction stir welded composite plates**

## 4. Results and Discussions

The friction stir welded composite plates are tested as per ASTM standards for ultimate tensile strength and hardness. The specimen taken for ultimate tensile strength is shown in figure 7. The values obtained after testing is narrated in table 2.

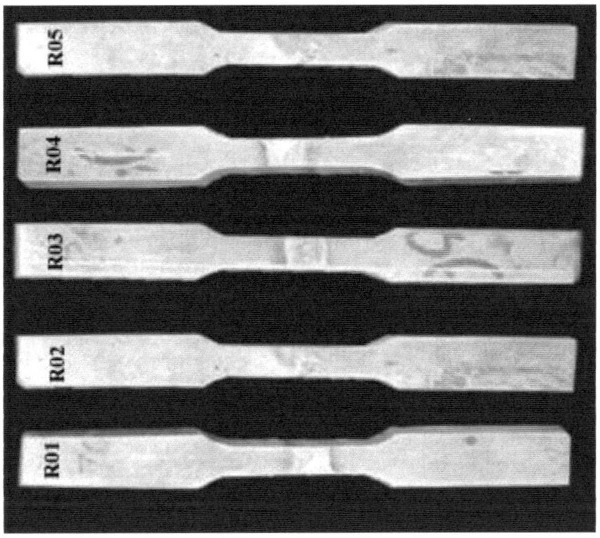


Figure 7 Specimen for ultimate tensile strength

Table 2 Ultimate tensile strength and hardness results for welded composites

| Sl.No | Tool Pin Profile | Ultimate tensile strength | Hardness |
|---|---|---|---|
| 1. | Square | 228 | 56 |
| 2. | Hexagon | 198 | 42 |
| 3. | Taper | 206 | 46 |
| 4. | Round | 212 | 48 |
| 5. | Threaded round | 214 | 52 |

The above results show the square shaped tool gives the best result in hardness and ultimate tensile strength. The hexagon will produce least among the five types of tools used for welding and there will be minor difference between the round shaped tool and threaded round tool. Taper shape pin profile results the moderate among all the five tools used in this research work.

## References:

[1] Mohsen hajizamani,,HamidrezaBaharvandi,"Fabrication and Studying the Mechanical Properties of A356 Alloy Reinforced With Al2O3-10% Vol.ZrO2 Nanoparticles through Stir Casting", Advances in Material Physics and Chemistry, pp 26-30 ,2011

[2] Yashpal,Sumankant,C.S. Jawalkar"Fabrication of aluminium metal matrix composites using particulate reinforcement",Materials today proceedings, vol-4,pp 2927-2936, 2017

[3] Rajeshkumar Gangaram Bhandre, Parshuram M. Sonawane "Preparation of Aluminium Matrix Composite by Using Stir Casting Method", International Journal of Engineering and Advanced Technology (IJEAT), vol-2,2013.

[4] YanpeiSong,ShuangxuBi, XiuqingLi"Microstructures and Properties of Recycled Composites Particle Reinforced Iron Matrix Functionally Graded Materials Fabricated by Centrifugal casting", pp360-366,Engineering,2010.

[5] Sourabh Gargatte* , Rahul R. Upadhye, Venkatesh S. Dandagi, Srikanth R. Desai, Bhimappa S. Waghamode, "Preparation & Characterization of Al-5083 Alloy Composites", Journal of Minerals and Materials Characterization and Engineering, pp 8-14, 2013

[6] Inampudi Narasimha Murthy, "Comparative Studies on Microstructure and Mechanical Properties of Granulated Blast Furnace Slag and Fly Ash reinforced AA2014 Composites", Journal of Minerals and Material Characterization and Engineering, pp8-16 , 2014.

[7] A. Ramesh, J. N. Prakash, A. S. Shiva Shankare Gowda and Sonnappa Appaiah "Comparison of the Mechanical Properties of AL6061/Albite and AL6061/Graphite Metal Matrix Composites", Journal of Minerals & Materials Characterization & Engineering, Vol. 8, No.2, pp 93-106, 2009.

[8] Yu Li, Qiu-lin, Fabrication and characterization of stir casting AA6061-31%$B_4C$ composite Structural and mechanical properties of aluminium based composites,Transactions of non ferrous metals society of china, vol 26,2304-2312,2016.

[9] Issac dinakaran,kumaravel kalaiselvan,Influence of rice husk ash particles on microstructure and tensile behavior of AA6061 aluminium matrix composites produces using friction stir processing. Composites communications, vol 3:42–46,2017.

[10] Rajakumar.S,Muralidharan.C , Balasubramanian.V, Influence of friction stir welding process and tool parameters on strengthproperties of AA7075-T6 aluminium alloy joints.Materials and Design, vo l32;535-549,2011.

[11] Arun kumar kadian,Pankaj biswas, Effect of tool pin profile on the material flow characteristics of AA6061,Materials today proceedings, vol-26 pp 382-392, 2017.

[12] Vijay ShivajiGadakh, Kumar Adepu,Heat generation model for taper cylindrical pin profile in FSW, j mater res technol vol2(4): 370–375,2013.

[13] Kumar A, Raju LS. Influence of tool pin profiles on friction stir welding of copper. Material Manufacturing ,vol 27(12):1414–18,2012.

[14] Sued M.K,Pons.D, Lavroff.D,Wong.E.H, Design features for bobbin friction stir welding tools: Development of a conceptual model linking the underlying physics to the production process,Materials and Design,2014, vol54 :632–643.

[15] Srinivasa Rao.M.S, Ravikumar.B.V.R, Experimental study on the effect of welding parameters and tool pin profiles on the IS:65032 aluminium alloy FSW jointsMaterials today proceedings, vol-4,pp 1394-1404, 2017

IOP Conf. Series: Materials Science and Engineering **247** (2017) 012016 doi:10.1088/1757-899X/247/1/012016

# Weldability of AA 5052 H32 aluminium alloy by TIG welding and FSW process – A comparative study

Shanavas S [1] and Edwin Raja Dhas J [2]
[1].Research Scholar, Department of Mechanical Engineering, Noorul Islam University, Tamil Nadu, INDIA.
[2].Deputy Director - Admissions, Noorul Islam University, Tamil Nadu, INDIA.

**Abstract**

Aluminium 5xxx series alloys are the strongest non-heat treatable aluminium alloy. Its application found in automotive components and body structures due to its good formability, good strength, high corrosion resistance, and weight savings. In the present work, the influence of Tungsten Inert Gas (TIG) welding parameters on the quality of weld on AA 5052 H32 aluminium alloy plates were analyzed and the mechanical characterization of the joint so produced was compared with Friction stir (FS) welded joint. The selected input variable parameters are welding current and inert gas flow rate. Other parameters such as welding speed and arc voltage were kept constant throughout the study, based on the response from several trial runs conducted. The quality of the weld is measured in terms of ultimate tensile strength. A double side V-butt joints were fabricated by double pass on one side to ensure maximum strength of TIG welded joints. Macro and microstructural examination were conducted for both welding process.

**Keywords:** Aluminum Alloy 5052 H32, Friction Stir Welding, TIG welding, Tensile Strength, Microstructure.

## 1. Introduction

AA 5052 H32 is cold-rolled aluminium alloy having very high corrosion resistance with an industrial and marine environment, good formability, and good weldability out of other aluminium alloys and thus has been used for automotive and marine structural applications. Friction stir welding (FSW), is a solid state joining process and its application found in various industries like aerospace, marine, automobile, etc. due to its high-quality joints [1, 2]. Since no fusion of material is involved, it avoids many weld defects caused by metallurgical reaction and distortion in fusion welding. It can weld both similar and dissimilar materials that cannot be weld by the conventional welding process. In FSW process the welding parameters such as tool pin profile, rotational speed, welding speed, tool tilt angle, etc. are responsible for quality joints [3-6].

Tungsten Inert Gas (TIG) welding is a fusion welding process widely used for joining ferrous and nonferrous metals. The process produces quality joints. Selection of filler rod, shielding gas flow rate, welding current, polarity of weld, arc voltage, and welding speed are some of the variables that affect the quality of the weld [7, 8]. These variables are dependent and change in one generally requires a change in others to produce a good weld. The proper selection and levels of significant parameters are critical for quality welds.

Adalarasan et al. [7] developed a Taguchi's L9 orthogonal array and grey relational analysis to optimize the welding parameters of TIG welded AA 6061 aluminium alloy. The process parameters such as arc voltage, current, welding speed, and gas flow rate were varied to find its effect on mechanical properties of the joints. The results concluded that the contribution of welding current and gas flow rate has significant effect on weld quality. Jose et al. [8] studied the influence of the welding parameters on the heat affected zone (HAZ) of aluminium structures. A response model is used to optimize the welding parameters to obtain a reduced HAZ. It was found that the greater temperature and lower cooling rate increased HAZ region. Sunil et al. [9] performed TIG welding on AA 7075-T6 aluminium alloy. Single V-butt joint is employed and the mechanical characteristics of the joint were determined experimentally and numerically and the results were compared.

Ahmed et al. [10] investigated the effect of welding speed on the tensile strength of the single V-butt TIG welded joint. Aluminium AA 6351 alloy is selected for experimentation with different bevel angle and bevel heights. It was found that tensile strength became higher with lower weld speed and at bevel angles between $30^0$ to $45^0$. Arun et al. [11] presented a work deals with the influence of TIG welding parameters in AA5083 aluminium alloy using factorial experimental design. Welding current and shielding gas flow rate were considered as the predominant input parameters. Ghazvinloo et al. [12] analyzed the effect of arc voltage, welding current and welding speed on mechanical properties of the joint of AA 6061 aluminium alloy. Result revealed that increase in heat input decreases the fatigue life whereas impact energy initially increased and then decreased.

Depending upon the welding process used, the strength of the weld differs for the same material. A comparative study on welding process for the same material is essential for the selection of the welding process. Several researchers compared TIG welded joints with FS welded joints for aluminium alloys to choose the right process [13-16]. It was found that FSW process is effective than TIG welded joints for most aluminium alloys. Quan et al. [17] proposed a double pass TIG butt welding on 2219-T8 aluminium alloy. The strength of the joint was enhanced and fracture occurred at the welded metal in the partially melted zone. From the survey, it was observed that a study of TIG welding and FSW process on AA 5052 H32 aluminium alloy was scarce. Hence an attempt was made in the work to find the effect of the welding current and inert gas flow rate on the tensile strength of the weld pool by TIG welding and the optimum result was compared with predetermined optimum FS welded joint. Macro and micro structures were examined for both welding process on the developed specimen and the reports were presented.

## 2. Experimentation

High strength aluminium magnesium alloy AA 5052-H32 cold rolled plates of size 100 mm × 50 mm × 6 mm were used for experiments. The chemical composition of the metal is shown in Table 1. Wire brushing is used for cleaning the adjoining surfaces for welding. The direction of welding was parallel to the rolling direction of the plate. A double side V-butt joints [Fig. 1] were fabricated by double pass on one side for TIG welding process. The process is performed using the AA 4043 filler rod of diameter 1.6 mm and using argon as shielding gas. The selected input variables for TIG welding are welding current and inert gas flow rate. Other parameters such as welding speed and arc voltage were kept constant.

Table 1
Chemical composition of AA 5052 H32 alloy

| Element | Cu | Mn | Mg | Si | Cr | Fe | Ti | Al |
|---|---|---|---|---|---|---|---|---|
| Wt. % | 0.01 | 0.08 | 2.33 | 0.14 | 0.18 | 0.29 | 0.02 | Bal. |

In FSW, the plates were welded in a single pass, using tapered square tool pin profile with a taper angle of $10^0$ and pin length of 5.7 mm (Fig. 2). Parameters such as tool tilt angle, rotational speed, and welding speed were kept optimum at $1.5^0$, 600 rpm, and 65 mm/min respectively. Photograph of the fabricated TIG and FS welded joints are shown in Fig. 3 and Fig. 4 respectively.

ASTM – E8 standard was followed for conducting the tensile test [18]. Tensile tests were carried out by UTM, FSA-M100 with 100 KN capacities. The results of experimental runs for various combinations of welding parameters are shown in Table 2. Microstructure examinations were carried out on the cross sections of the welded joints. The specimens for examination is initially cut by EDM, and progressively grind using different grades of emery paper, then polished using diamond paste, and finally etched with modified keller reagent.

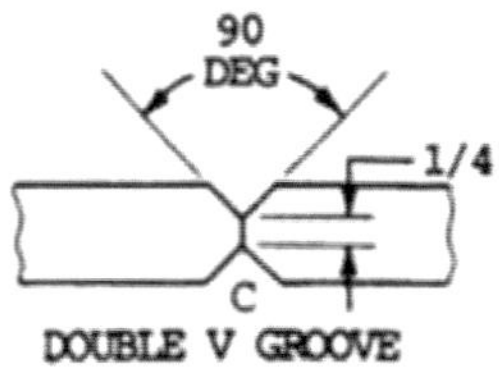


Fig. 1. Joint design for TIG welding

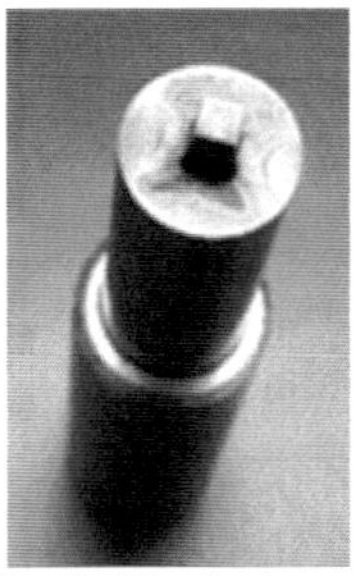

Fig. 2. Tool pin used for FSW

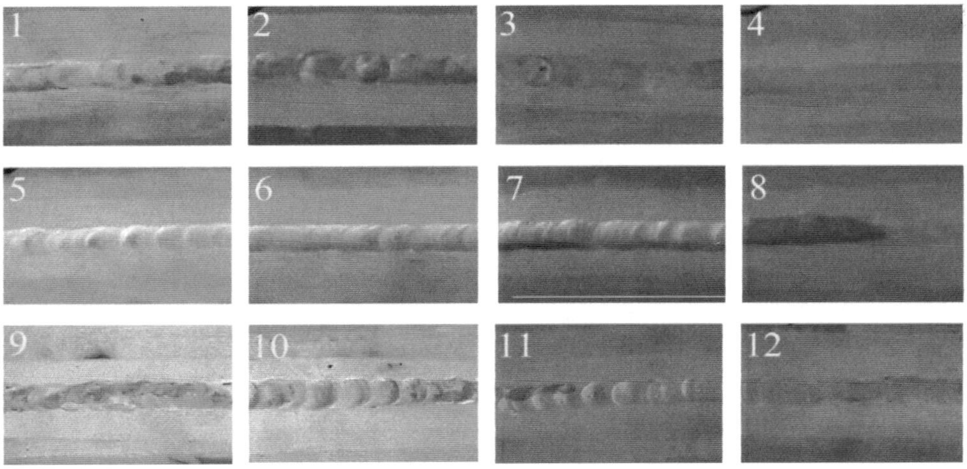


Fig. 3. Photograph of the joints fabricated by TIG welded process.

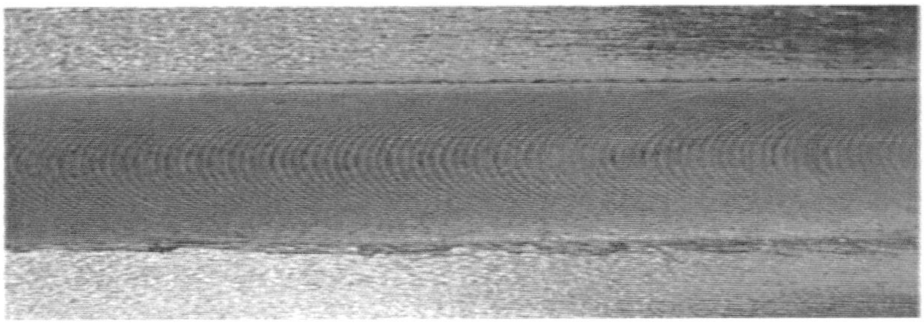

Fig. 4. Photograph of the joint fabricated by FSW process

Table 2

IOP Conf. Series: Materials Science and Engineering **247** (2017) 012016 doi:10.1088/1757-899X/247/1/012016

Combination of welding parameters and the observed response

| Trials | Input parameters | | | | Response | | | |
|---|---|---|---|---|---|---|---|---|
| | Weld Current (A) | Gas flow rate (lpm) | Arc Voltage (V) | Weld speed (mm/min) | YS (Mpa) | UTS (MPa) | % E at Yield | % E at Break |
| 1 | 160 | 9 | 20 | 100 | 100.39 | 125.53 | 2.94 | 18.22 |
| 2 | 170 | 9 | 20 | 100 | 118.22 | 168.64 | 8.62 | 16.94 |
| 3 | 180 | 9 | 20 | 100 | 101.03 | 158.42 | 5.00 | 24.22 |
| 4 | 190 | 9 | 20 | 100 | 90.00 | 146.69 | 3.63 | 10.93 |
| 5 | 160 | 11 | 20 | 100 | 83.36 | 105.25 | 5.34 | 6.90 |
| 6 | 170 | 11 | 20 | 100 | 106.58 | 133.28 | 5.13 | 7.12 |
| 7 | 180 | 11 | 20 | 100 | 101.06 | 186.58 | 5.20 | 28.38 |
| 8 | 190 | 11 | 20 | 100 | 92.94 | 154.78 | 3.20 | 23.16 |
| 9 | 160 | 13 | 20 | 100 | 80.75 | 147.36 | 3.48 | 15.01 |
| 10 | 170 | 13 | 20 | 100 | 79.64 | 150.14 | 3.57 | 25.26 |
| 11 | 180 | 13 | 20 | 100 | 86.89 | 155.31 | 3.55 | 24.98 |
| 12 | 190 | 13 | 20 | 100 | 67.69 | 127.36 | 3.03 | 14.40 |

## 3. Results and discussion

### 3.1. Tensile test

Tensile tests were carried out on TIG welded plates. As per ASTM – E8 standard the specimens were prepared from the TIG welded plates. Photograph of the tensile tested specimens is displayed in Fig. 5. Interaction effects of welding current and inert gas flow rate on ultimate tensile strength (UTS) is shown in Fig. 6. The results reveal that for various inert gas flow rate, the increase in welding current initially increase the tensile strength, reaches a maximum value and then decrease. The increase in welding current improve the metal penetration during the TIG welding process, thereby improves weld properties. The lower value of current resulted in reduced fusion zone leads to poor joint strength. The higher and lower welding current lead to poor bonding due to high heat generation and low heat generation respectively. The optimum heat generation creates a quality weld. The result also shows that a moderate level of shielding gas flow rate (11 lpm) was necessary for better strength of joints. A lower level of gas flow rate cannot be recommended to produce better response. The maximum ultimate strength of the joint fabricated by TIG welding experimental runs is 186.58 MPa and that produced by FSW is 202.04 MPa. It is observed that the FSW process can produce a better joint than TIG welding process with AA 5052 H32 aluminium alloys.

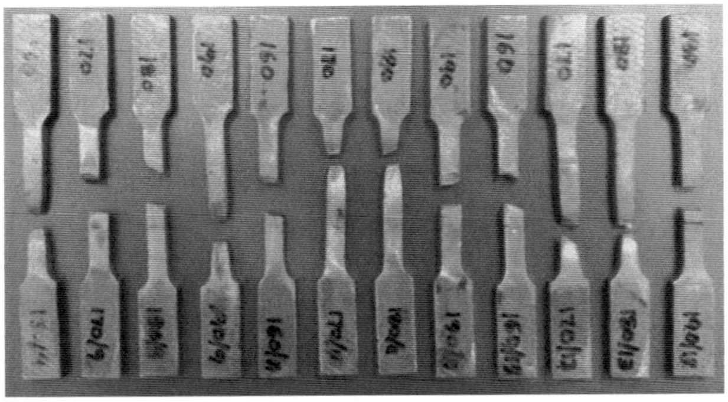

Fig. 5. Photograph of the tensile tested specimens

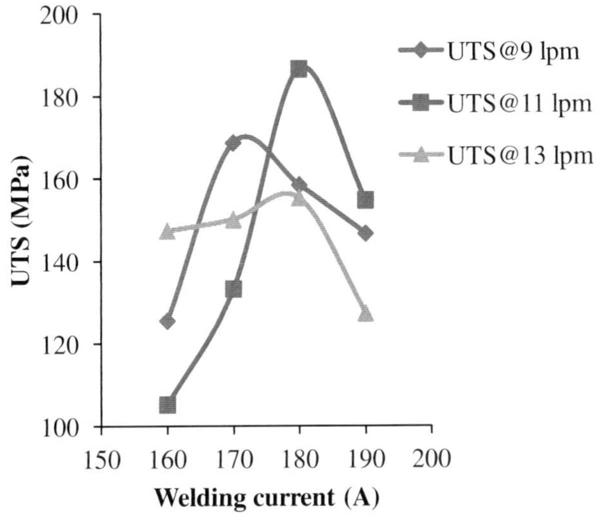


Fig. 6. Interaction effects of welding current and inert gas flow rate on UTS

### 3.2. Macro and microstructures

The optical macrograph and optical microstructure of the both TIG and FS welded joints obtained with an optimal setting of parameters is shown in Fig. 7 and Fig. 8 respectively. Defects such as voids, cracks, and unbonded zones are not observed in and around the weld area for weld fabricated by both processes (Fig. 7). Compared to the parent metal, grain growth was visualized at the weld area (fusion zone) in TIG welds (Fig. 8b). Grain growth is due to the elevated temperature during fusion welding. Coarsening of grain reduces the strength in the fusion zone. Dendritic structure appeared at fusion zone was due to the faster heating and cooling of the weld metal. Fig. 8c shows a clear visualization of dendritic spacing at weld site. At heat hazard region (HAZ) the dendritic structure gets reduced (Fig. 8d). Relatively fine dendritic structure was observed at HAZ. Fine grain structure was observed at SZ of the plates welded by FSW due to recrystallization at stir zone (Fig. 8e).

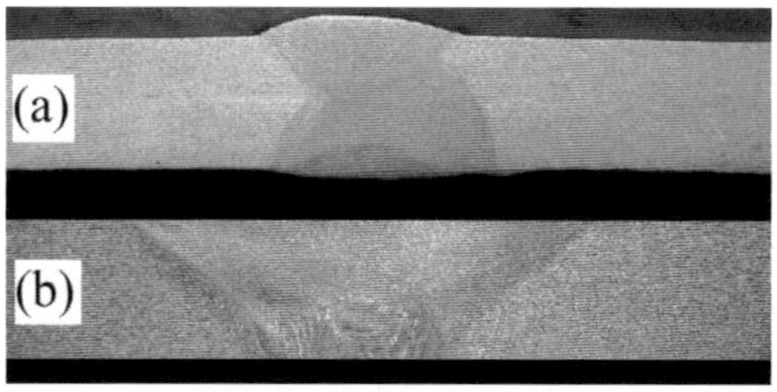


Fig. 7. Cross sectional macrograph of the welded specimen produced by: (a) TIG welding, (b) FSW

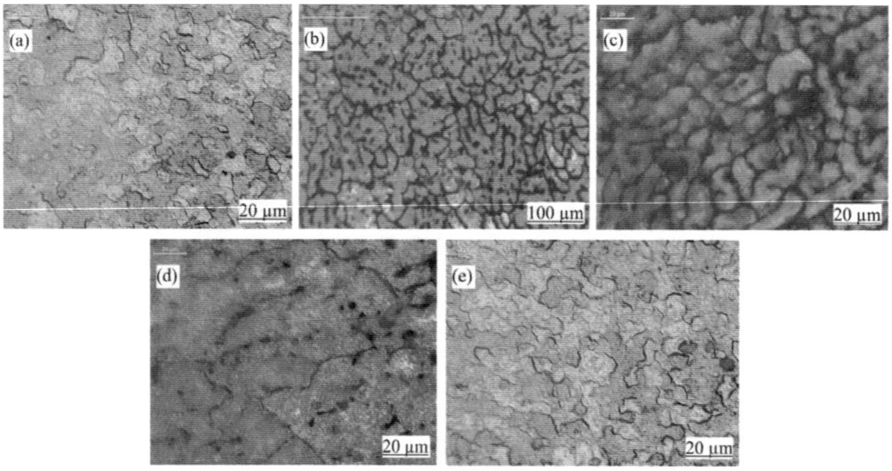


Fig. 8. Optical micrographs of (a) parent metal, (b) & (c) weld site of TIG welded plate at different magnification, (d) HAZ of TIG welded plate and (e) stir zone of FS welded plate.

## 4. Conclusions

The present work made a comparative study on the metallographic and mechanical properties of the joints obtained by the TIG welding and FSW processes with AA 5052 H32 aluminium alloy. The following conclusions can be drawn:

1) The joint fabricated by the FSW process were found to posses better metallographic and mechanical properties than those fabricated by using the TIG welding process.

2) The welding current and inert gas flow rate play a significant role in determining the quality of the TIG welded joint.

3) In TIG welding, the joint fabricated by a welding current of 180 Ampere, inert gas flow rate of 11 lpm, welding speed of 100 mm/min, and an arc voltage of 20 V exhibited superior tensile properties compared to other joints.

## References

[1] A Grimm, S Schulze, A Silva, G Gobel, J Standfuss, B Brenner, E Beyer, U Fussel. Friction stir welding of light metals for industrial applications. Materials Today: Proceedings. 2015;2S:169 – S178.
[2] A Anand, A Khajuria. Welding processes in marine application: A review. International Journal of Mechanical Engineering and Robotics Research. 2015;2: 215-225.
[3] Raza Moshwan, Farazila Yusof, M A Hassan, S M Rahmat. Effect of tool rotational speed on force generation, microstructure and mechanical properties of friction stir welded AA 5052-O alloy. Materials and Design. 2015; 66:118–128.
[4] Leonardo N Tufaro, Ivan Manzoni, Hernan G Svoboda. Effect of Heat Input on AA5052 Friction Stir Welds Characteristics. Procedia Materials Science. 2015;8:914 – 923.
[5] S M Bayazid, H Farhangi, A Ghahramani. Effect of Pin Profile on Defects of Friction Stir Welded 7075 Aluminum Alloy. Procedia Materials Science. 2015;11:12 – 16.
[6] S Ugender, A Kumar, A Somi Reddy. Experimental investigation of tool geometry on mechanical properties of friction stir welding of AA 2014 aluminium alloy. Procedia Materials Science. 2014;5:824-831.
[7] R. Adalarasan, M. Santhanakumar. Parameter design in fusion welding of AA 6061 aluminium alloy using desirability grey relational analysis (DGRA) method. J. Inst. Eng. India Ser. C. 2015: 96:57-63.
[8] Jose L. Meseguer-Valdenebro, Eusebio J. Martinez-Conesa, Jose Serna, Antonio Portoles. Influence of the welding parameters on the heat affected zone for aluminum welding. Thermal Science. 2016 : 20 : 643-653.
[9] Sunil M. Pawar, Vivek V. Kulkarni. Mechanical characterization of aluminium alloys for TIG welding-experimental and modeling studies. International Research Journal of Engineering and Technology. 2015 : 2 : 1086-1089.
[10] Ahmed Khalid Hussain, Abdul Lateef, Mohd. Javed, T. Pramesh. Influence of welding speed on tensile strength of welded joint in TIG welding process. International journal of applied engineering research, 2010 : 3 : 518-527.
[11] Arun Narayanan, Cijo Mathew, Vinod Yeldo Baby, Joby Joseph. Influence of gas tungsten arc welding parameters in aluminium 5083 alloy. International Journal of Engineering Science and Innovative Technology. 2013 : 2 : 270-277.
[12] H. R. Ghazvinloo, A. Honarbakhsh-Raouf, N.Shadfar. Effect of arc voltage, welding current and welding speed on fatigue life, impact energy and bead penetration of AA6061 joints produced by robotic MIG welding. Indian Journal of Science and Technology. 2010 : 3 : 1-7.
[13] Gurmeet Singh, Amardeep S. Kang, Kulwant Singh, Jagtar Singh. Experimental comparison of friction stir welding process and TIG welding process for 6082-T6 Aluminium alloy. materials today: proceedings. 2017 : 4 : 3590–3600.
[14] Baiju sasidharan, K. P. Narayanan, R. S. Prakash. Tensile and microstructural characterestics of dcsp tig welded and friction stir welded aa2219 aluminium alloy. International journal of design and manufacturing Technology. 2014 : 5 : 121-129.
[15] Zhen-bo HE. Yong-yi PENG. Zhi-min YIN. Xue-feng LEI. Comparison of FSW and TIG welded joints in Al-Mg-Mn-Sc-Zr alloy plates. Transactions of Nonferrous Metals Society of China. 2011 : 21: 1685-1691.
[16] A. Cabello Munoz, G. Ruckert, B. Huneau, X. Sauvage, S. Marya. Comparison of TIG welded and friction stir welded Al–4.5Mg–0.26Sc alloy Journal of Materials Processing Technology. 2008 : 197: 337–343.
[17] Quan LI. Ai-ping WU. Yue ZHAO. Guo-qing WANG. Dong-yang YAN. Fracture behavior of double-pass TIG welded 2219-T8 aluminum alloy joints under transverse tensile test. Transactions of Nonferrous Metals Society of China. 2015 : 25 : 1794-1803.

[18] ASTM E8/E8M-09. Standard test methods for tension testing of metallic materials1. ASTM International; 2010.

# Design of Smooth Ramp Feedrate for Machining Complex NURBS Paths

M. Sekar*
Department of Mechanical Engineering
GMR Institute of Technology
Rajam, India
mailtosekar@gmail.com

B. Suresha
Department of Mechanical Engineering
National Institute of Engineering
Mysuru, India
sureshab2004@yahoo.co.in

I. Kantharaj
Department of Mechanical Engineering
Karunya University
Coimbatore, India
kantharaj@karunya.edu

*Abstract:*
NURBS curves are widely adopted in CAD for the purpose of aesthetics and became functional requirements. In machine tool controllers, the position commands of NURBS tool paths are obtained from incremental parametric values. For technical reasons, the tool must be commanded to reach the parameter value corresponding to the end point of the free-form curve; however this is not guaranteed by the Taylor's expansion. This aspect restricts the implementation of a free-form parameteric interpolator for NURBS machining in actual machining practice. In this paper a non-uniform parametric algorithm is proposed to find the ramp increments. In this approach the difference between the computed and the end parameter values is represented as an area of triangle in the time domain with its base equal to the integer values of sampling interval set by the machine controller. Results show that the proposed simulator designed can reach the end parameter value exactly. The novel ramp feedrate method suggested in this paper is also useful for multi-pass cutting in CNC machine.

*Keywords:* Parametric NURBS, CNC controller, smooth ramp profile, SIMULINK, Real-Time Workshop.

## INTRODUCTION

In CNC machines controller outputs the position commands for a parametric NURBS curve using approximation methods. Computer aided manufacturing (CAM) software employ techniques that divides the NURBS into linear G codes segments. However such techniques result in constraints on resources, fluctuating feedrate leading to stability problems and further polishing work to achieve the desired surface finish [1]. In addition the coordinate points computed by the interpolator may not actually represent the required NURBS curve designed in CAD.

The real time NURBS interpolator introduced by Shiptalni et.al has gained wide acceptance among NURBS implementations as the CAD geometry and can directly be employed for interpolation while CNC machining of components designed with NURBS [2]. In particular the parametric incremental interpolators proposed by [3-6] does not need constant increments of chord length. One of the unsolved issues in

* Corresponding author

parametric curve machining is to reach the end coordinate value irrespective of the given feedrate and control parameters. Computing parameter value from a coordinate is not feasible with NURBS as closed loop mathematic equations are not yet developed. A mismatch between the manufactured product and the NURBS design will be inevitable if the end parameter value corresponding to the final point NURBS is not met. In this article the solution to this problem is solved by representing the difference *(du)* between the computed parameter and the end parameter as an area of triangle in the time domain. Here *du* the area of triangle has the integer multiple of controller sampling intervals.

After the brief overview of NURBS is outlined, the Taylor's series for real-time parametric interpolation is introduced. The problems pertaining to the reach the end parameter value is then discussed. The solution to the above problem is proposed in the next section followed with details of the development of simulator. Information on development of SIMULINK blocks and implementation in real-time workshop is presented in this article.

FEEDRATE DESIGN FOR NURBS TOOL PATHS

A NURBS curve is a piecewise parametric curve represented by a rational polynomial [7]

$$\mathbf{C}(u)=\frac{\sum_{i=0}^{n}N_{i,p}(u)\,w_i\mathbf{P}_i}{\sum_{i=0}^{n}N_{i,p}(u)\,w_i} \quad \text{with } 0\le u\le 1 \text{ and } w_i>0 \tag{1}$$

where the control polygon consists of $P_i$ control points, weight vectors with values $w_i$ and the $N_{i,p}(u)$ are the $p^{th}$ degree B-spline basis functions defined on non-uniform knot vector $U$

$$U=\{0,\ldots,0,u_{p+1},\ldots\ldots,u_{m-p-1},1,\ldots,1\} \tag{2}$$

The basis function of the spline with degree $p$ can be recursively written as

$$N_{i,0}(u)=\begin{cases}1 & if\ u_i\le u\le u_{i+1}\\ 0 & otherwise\end{cases} \tag{3}$$

$$N_{i,p}(u)=\frac{u-u_i}{u_{i+p}-u_i}N_{i,p-1}(u)+\frac{u_{i+p+1}-u}{u_{i+p+1}-u_{i+1}}N_{i+1,p-1}(u) \tag{4}$$

while the relation between m, n and p can be represented as

$$m=n+p+1. \tag{5}$$

The derivative of the basis function is widely used for determining the increments and is given by

$$N'_{i,p}(u)=\frac{p_i}{u_{i+p}-u_i}N_{i,p-1}(u)+\frac{p}{u_{i+p+1}-u_{i+1}}N_{i+1,p-1}(u) \tag{6}$$

*A. Real time Interpolator for CNC machines*

NURBS interpolation can result in the uniform of non-uniform increment to the parameter values. The values obtained are substituted in Eqn.1 to determine the corresponding command position. In contrast to the conventional CNCs, NURBS position commands are obtained by considering both parametric domain (u) and time domain (t). However there exists no mathematical functional relationship between the u and t domains. Hence the real time interpolator use numerical integration methods to find the next immediate parameter for every sampling period. For a NURBS curve represented by Eqns. 1-6, the incremental change of parameter value in time domain is obtained from the command velocity *V(u)* as follows:

$$V(u)=\left\|\frac{dC(u)}{dt}\right\| \quad (7)$$

Differentiating the parameter u with respect to time

$$\frac{du}{dt}=\frac{V(u)}{\left\|\frac{dC(u)}{du}\right\|}, \quad (8)$$

Applying Taylor's series for every sampling interval $T_s$, $t=kT_s$ with $k$ as the number of samples the next incremental value to parameter can be computed as

$$u_{k+1}=u_k+\Delta T\,\dot{u}\big|_{t=t_k}+\frac{(\Delta T)^2}{2}\ddot{u}\big|_{t=t_k}+\frac{(\Delta T)^3}{6}\dddot{u}\big|_{t=t_k}+\ldots \quad (9)$$

$$\dot{u}=\frac{V}{\sigma} \quad (10)$$

$$\ddot{u}=\frac{\sigma\dot{V}-\sigma' V\dot{u}}{\sigma^2}=\frac{\dot{V}}{\sigma}-\frac{\sigma' V\dot{u}}{\sigma^2} \quad (11)$$

$$\dddot{u}=\frac{\ddot{V}}{\sigma}-\frac{3\sigma' V\ddot{u}}{\sigma^2}-\frac{\sigma'' V\dot{u}^2}{\sigma^2} \quad (12)$$

On revisiting the Taylors's equation for NURBS interpolation and the recursive equations to compute the derivatives of parameter in time domain, the first and second order derivatives of instantaneous parameters are necessary [8].

*B. Investigation of Parametric Interpolator*

In real-time CNC controller design, the X and Y positions of a parametric curve do not use the HOTs for computation because of the recursive polynomial functions of NURBS. A continuous feedrate within the limits of acceleration and jerk is preferred for better machine dynamics [9]. Smooth feedrate profile results in a coordinated motion with reduced tracking errors due to rapid and accurate execution of low-jerk trajectory. Though there are many practical issues in the implementation of a real-time parametric interpolator, this paper discusses one of the key issues regarding interpolating exactly until the end parameter value of the given curve.

It is a standard practice to normalize the knot vector from *0-1*. In CNC controller the initial parameter value equal to *0* is substituted in equation *1* to find the coordinates. To generate the coordinate position for the next sampling time, $u_{i+1}$ is required which can be computed by implementing Eqn. 9 in the CNC controller. The process is iterated with $k$ times until the parameter reaches the maximum value of $u=1$. Since the Taylor's interpolator computes increments to the parameter non-linearly, the end parameter value i.e. $u=1$ may not reached at all times. Considering the geometry of the NURBS curve, large distances may be left to cover even for small incremental change in parameter values. This results in a machining process where the final coordinates are left un-machined leaving an undercut or overcut in the components. A reduction in surface accuracy, dimensional deviation and assembly problems are inherent in this method.

*C. Proposed Solution for Parametric Interpolator with smooth ramps*

In the proposed non-uniform parametric increment algorithm, the change ($du$) between the computed $u_{end\_Taylor}$ and $u=1$, is represented as an area of triangle in the time domain (Fig. 1). The objective is to find cumulative sum of area at each sampling interval. The number of sampling intervals represents the width ($B$) of the triangle and can be provided as an input value. The instantaneous value of '$h$' can be found from similar triangles, where '*H*' is the height of the triangle. Referring to Fig.1, the difference between the final parameter $u=1$ and the starting parameter is mentioned as *du*. In time domain *du* is expressed as the area of triangle with the base equal to *n* multiples of controller sampling time. A triangular profile is chosen because the increments in *u* can be increased and decreased smoothly in the parametric domain.

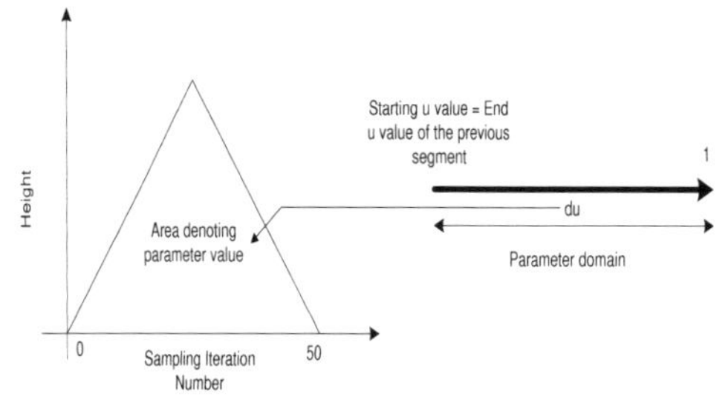


Fig. 1. Remaining parameter *u* represented in time domain.

The area of the triangle in Fig. 2 represents the *du*. At each time interval *b* of total time *B*, area can be obtained based on similar triangles *0-b-h* and *0-B/2-H*. The parameter value corresponding to area at *b* can be obtained as the sum of starting parameter (*1-du*) and the area $a_b$.

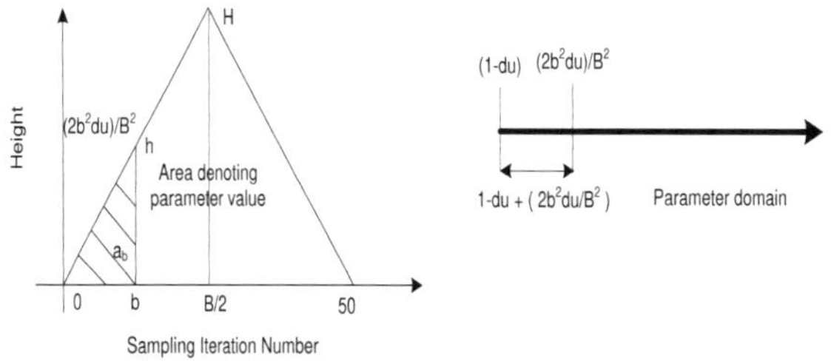


Fig. 2. Remaining *u* during first half of travel.

$$u_b = (1 - du) + a_b \text{ where } a_b = 0.5bh \tag{13}$$

Using similar triangles, we get,

$$u = (1 - du) + \frac{b^2 2du}{B^2} \quad (if\ b \le B/2) \tag{14}$$

The increase in *b* results in smooth increase of *u* value. Similarly for the other condition ($b>B/2$) the following formula can easily be derived.

$$u = 1 - \frac{b^2 2du}{B^2} \quad (if\ b > B/2) \tag{15}$$

*D. Implementation in SIMULINK and Real-Time Workshop*

MATLAB provides a power tool namely the Real-Time Workshop which extends the capabilities of both SIMULINK and MATLAB functionalities to a target hardware [10]. Generally the target hardware is not compatible to MATLAB *m* files and the proposed CNC architecture requires special blocks which are not available in the SIMULINK. Therefore customized codes are written in C to develop the interpolator block set. In this paper details regarding the real-time interpolator block are discussed.

The algorithm consisting of Eqns. 1-15 are implemented as level 2 SIMULINK S function. The Real-Time Workshop of MATLAB has the target link compiler to convert the MATLAB code into executable. With the automatic code generation tailored for a variety of target platforms, these codes can easily be deployed with RTW. RTW compatible SIMULINK blocks support platform independency and promote rapid development and deployment of systems. In addition the integration with MATLAB provides sharing of data to the workspace so as to enable plotting and carry out other analysis in MATLAB command prompt.

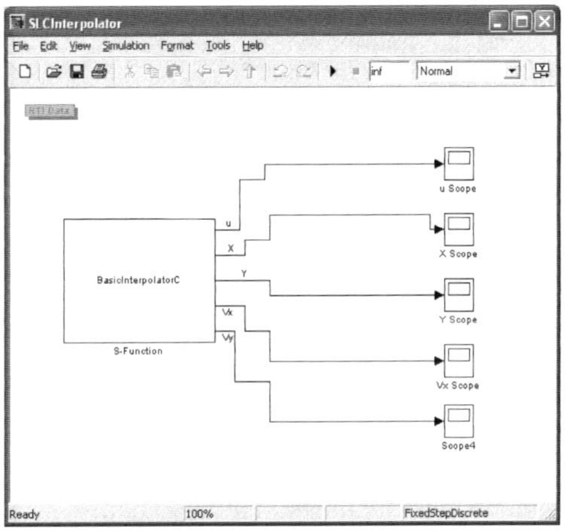


Fig. 3. Interpolator blocks using SIMULINK S2 function.

Fig. 3 shows the interpolator SIMULINK block implemented for the purpose of simulation and real-time execution in a suitable hardware target. The code generation report from the RTW has a list of various files generated to aid the real-time implementation of the SIMULINK block in a hardware platform. The files include source files, header files, data type definition files and object codes. The simulator consists of scopes and functions to export the computed values to MATLAB workspace for easy plotting. To enable diagnosis of the algorithm in real-time, global variables compatible to the hardware platform are defined. The corresponding header file of the target hardware is also essential for the successful uploading of the object codes into the controller memory.

*E. Testing the block in SIMULINK*

For testing of the interpolator block and the outputs, a circle is modeled as a parametric NURBS curve. Following figures show the outputs obtained from the interpolator SIMULINK block. In real-time implementation the coordinate data is sent to X, Y stage loop controllers to generate the required profile. The data corresponding to the outputs are also exported to workspace for the purpose of analysis. Fig. 4 shows the plot constructed using the data from the workspace and the Figs 5. a-c show the scope graphs taken from SIMULINK model.

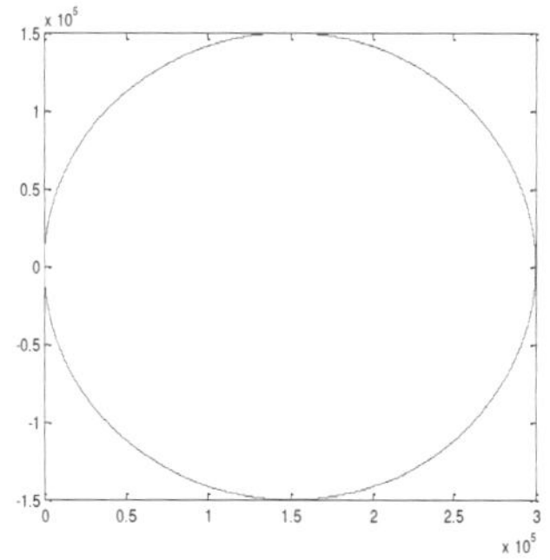


Fig. 4. Parametric circle profile generated from workspace data.

(a)

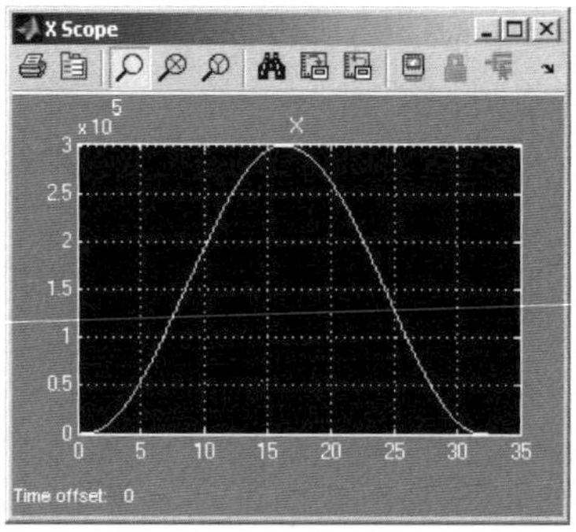


(b)

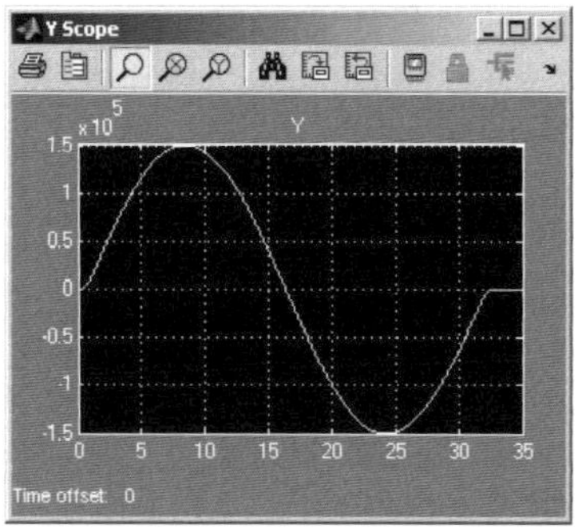


(c)

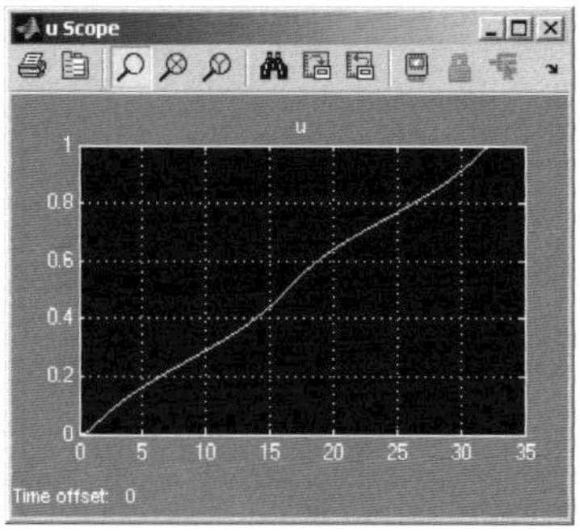


Fig. 5 a, b, c. Outputs from SIMULINK Interpolator block.
(Coordinates x, y, and parameter u plotted against time, Coordinates are in micro meters)

RESULTS AND DISCUSSION

Fig. 6 shows the sample NURBS parametric curve used to test the proposed algorithm with RTW generic target. The control points, weight vectors and knot vectors are listed in Table 1. The given curve has a degree of 2 so that second Taylor's approximation can be used in CNC interpolation except for the smooth ramping at the end. The important simulation parameters for the controller are shown in Table. 2. A sampling time of 10 *milli* seconds has been selected for the present simulation purpose. If the curve involves higher degree number of recursive iterations will increase and a small sampling time may not be enough to compute the coordinate points in real time. The developed code has features to change the sampling time after testing the entire code in a real-time system.

Table 1: NURBS data

| X(μm) | Y(μm) | W | U |
|---|---|---|---|
| 10000 | 0 | 1 | 0,0,0 |
| 20000 | 20000 | 1 | 0.2 |
| 12000 | 8000 | 1 | 0.4 |
| 10000 | 20000 | 1 | 0.6 |
| 8000 | 8000 | 1 | 0.8 |
| 0 | 20000 | 1 | 1 |
| 10000 | 0 | 1 | 1,1 |

Table 2: Simulation parameters

| | |
|---|---|
| Degree of the curve | 2 |
| No. of derivatives | 2 |
| Sampling time | 0.01 sec |
| Feedrate limit | 25000 mm/sec |
| Acceleration limit | 5000 $\mu m/sec^2$ |
| Jerk limit | 5000 $\mu m/sec^3$ |
| Chord error | 0.5μm |

With the initial feedrate of $V=1.5\times10^4$ *μm/sec* the Taylor's interpolator reaches the parameter value *0.9984*. Non-uniform increments based on the remaining *u* *(du=0.0016)* are computed using the ramping algorithm. The general interpolator has no inherent feedrate planning and hence cannot be directly implemented in the controlled since the controller cannot start with the command feedrate from the rest position. It is assumed that the solution for this problem exists and the feedrate schedules are available for the curve as shown in Fig. 7.

IOP Conf. Series: Materials Science and Engineering **247** (2017) 012017 doi:10.1088/1757-899X/247/1/012017

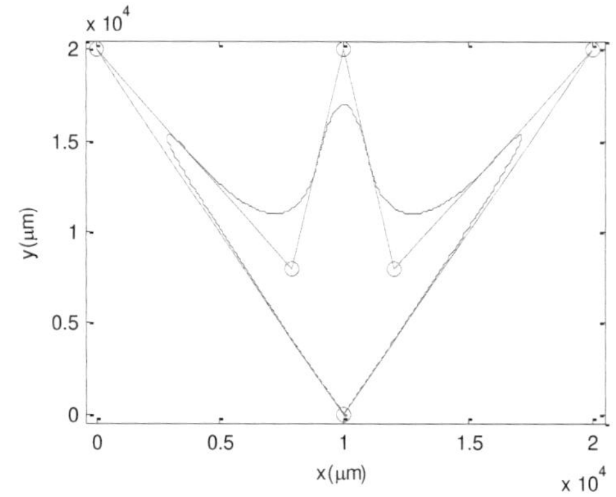


Fig. 6. Parametric NURBS curve

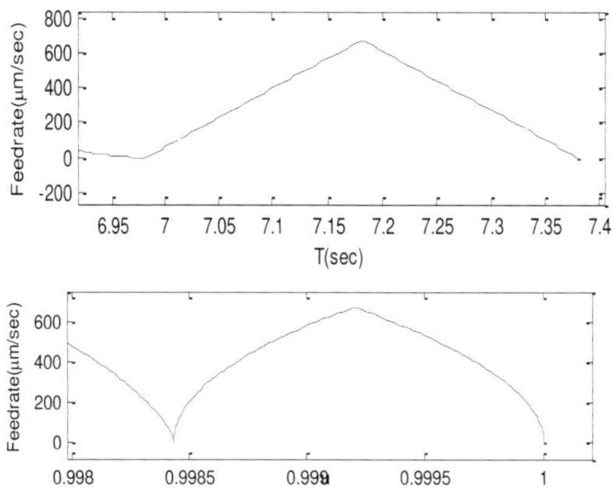


Fig. 7. Ramping algorithm in time and parameter domains

Table 3: Comparison of NURBS curve data

| u | X(μm) | Y(μm) |
|---|---|---|
| 0.9984 | 0 | 1 |
| 1 | 10000 | 0 |
| Δu=0.0016 | Δx=151 | Δy=302 |

It can be noticed if the algorithm is not used, the coordinate point corresponding to $u=1$ may never be reached and a distance of $\Delta x$ and $\Delta y$ will be left unvisited by the machine tool (Table. 3). This may result in poor surface finish and may even lead to broken tool when high axial depth of cuts is employed. Moreover, a sharp profile will not be machined accurately and the desired curvature may get altered. Though this can be overcome with a shorter sampling time, reduction of sampling time involves complexity in real time calculations of NURBS coordinate. In addition, the CNC controller also must allocate time to compute control loops of the individual motors. It also requires time slice to process the feedback data in real-time. Hence the proposed method is practically suitable for real time implementations.

## CONCLUSIONS

A novel algorithm for the smooth reaching of the end parametric value in real-time interpolation of a NURBS parametric curve is proposed. The proposed algorithm is integrated to a CNC architecture using SIMULINK and simulation results are shown in this paper. Results show that with the use of proposed algorithm the parameter value can be reached exactly. In addition, the design of CNC interpolator as SIMULINK block with RTW permits rapid development and deployment of the CNC architecture to a variety of hardware platforms. The developed interpolator SIMULINK blocks can be used in both simulation and real-time environments.

## REFERENCES

[1] M. Sekar, J. Srinivas, N. Kang and S.H. Yang, "An Integrated Approach to the Improvement of Stability Lobes," Int.. J. of Precision Engineering and Manufacturing, Vol. 9, pp: 83-85, 2008.

[2] M. Shiptalni, Y. Koren, and C. C. Lo, "Real-time curve interpolators.", Computer Aided Design, Vol. 26, No. 11, pp. 832-838, 1994.

[3] M. Y. Cheng, M. C. Tsai and J. C. Kuo, "Real-time NURBS command generators for CNC servo controllers." International Journal of Machine Tool and Manufacture, Vol. 42, No. 7, pp. 801-813, 2002.

[4] S. S. Yeh and P. L. Hsu, "Adaptive-feedrate interpolation for parametric curves with a confined chord error." Computer-Aided Design, Vol. 34, No. 3, pp. 229-237, 2002.

[5] M. Sekar, S. H. Kweon, D. M. Lee, S. H Yang, "Parameteric NURBS curve Inerpolators : A review." International Journal of Precision Engineering and Manufacturing, Vol. 9, No. 2, pp. 84-92, 2008.

[6] Yong, T., Narayanasami, R. "A parametric interpolator with confined chord error, acceleration and deceleration for NC machining", Computer-Aided Design, Vol.35, pp.1249–1259, 2003.

[7] L. Piegl, and W. Tiller,The NURBS book, Springer, 1997.

[8] Kevin M. Nittler and Rida T. Farouki, "Efficient high-speed cornering motions based on continuously-variable feedrates", International Journal of Advanced Manufacturing Technology, Vol.37, No. 5 pp. 545–552, 2016.

[9] Zhen-yuan Ji, De-ning Song, Jian-wei Ma, Guo-qing Hu and Wei-wei Su, "A NURBS interpolator with constant speed at feedrate-sensitive regions under drive and contour-error constraints", International Journal of Machine Tools & Manufacture, Vol. 116, pp. 1-17, 2017.

[10] MATLAB and SIMULINK manuals, 2008.